The terms and circumstances
of human existence
can be expected to
change radically during the
next human lifespan.
Science, mathematics, and
technology will be
at the center of
that change—causing it,
shaping it,
responding to it. Therefore, they
will be essential to
the education of
today's children for
tomorrow's world.

What should
the substance and character
of such education be?

The purpose of this
report is to propose
an answer to that question.

SCIENCE
FOR ALL AMERICANS

A PROJECT 2061 REPORT ON
LITERACY GOALS IN SCIENCE,
MATHEMATICS,
AND
TECHNOLOGY

AMERICAN ASSOCIATION
FOR THE ADVANCEMENT OF SCIENCE
1989

Founded in 1848, the American Association for the Advancement of Science is the world's leading general scientific society, with more than 132,000 individual members and nearly 300 affiliated scientific and engineering societies and academies of science. The AAAS engages in a variety of activities to advance science and human progress. To help meet these goals, the AAAS has a diversified agenda of programs bearing on science and technology policy; the responsibilities and human rights of scientists; intergovernmental relations in science; the public's understanding of science; science education; international cooperation in science and engineering; and opportunities in science and engineering for women, minorities, and the disabled. The AAAS also publishes *Science,* a weekly journal for professionals, and *Science Books & Films,* a review magazine for schools and libraries.

ISBN 0-87168-341-5

AAAS Publication 89-01S

Library of Congress Catalog Card Number: 88-36252

© 1989 by the American Association for the Advancement of Science, Inc., 1333 H Street NW, Washington, D.C. 20005

Printed in the United States of America

CONTENTS

ACKNOWLEDGMENTS

Recalling an exchange between Alice and the Cheshire Cat in *Alice in Wonderland* and reflecting on education, James Bryant Conant once said that if you don't know where you want to go, any map will do. On behalf of the Board of Directors of the American Association for the Advancement of Science, I wish to express our gratitude for the willingness of the National Council on Science and Technology Education to undertake the enormously difficult task of helping the nation decide where it wants to go in science, mathematics, and technology education. Not only did the national council—under the energetic, informed, and thoughtful leadership of William Baker and Margaret MacVicar—approach its assignment with verve and dedication, but it produced what we believe is a landmark report. The report was submitted to the AAAS board of directors by the national council as the culmination of a three-year study involving hundreds of scientists, engineers, and educators. After careful study, the board unanimously accepted it.

The board wishes to acknowledge with special thanks the central role played by the members of the five Project 2061 Phase I scientific panels. (See Appendix A for a complete listing of the names and affiliations of the panel members and of all other participants in this study.) The five panels met frequently over a period of two years to address the question of what science, mathematics, and technology students should understand, and submitted reports to the national council stating their conclusions. In preparing *Science for All Americans,* the national council drew heavily on the panel reports but did not try to duplicate or summarize them. The panel reports, available from the American Association for the Advancement of Science as companion volumes to this report, consist of:

Biological and Health Sciences, by Mary Clark

Mathematics, by David Blackwell and Leon Henkin

Physical and Information Sciences and Engineering, by George Bugliarello

Social and Behavioral Sciences, by Mortimer Appley and Winifred B. Maher

Technology, by James Johnson

We wish to thank the many other people (see Appendix A) who contributed so generously to Phase I by providing useful suggestions and criticisms. The national council, the scientific panels, and the project staff were able to call on expert consultants at crucial points in their work for help in clarifying difficult issues and for advice on technical matters outside their own fields of competence. In addition, scientists, mathematicians, engineers, and educators reviewed various drafts of the reports and provided suggestions for improvement. The help of those individuals was invaluable, but this fact should not be taken to imply that they endorse the recommendations or that they are in any way responsible for any shortcomings.

Special thanks go to the staff of Project 2061. F. James Rutherford, who conceived and designed the project and serves as its director, and Andrew Ahlgren, the associate project director, prepared successive drafts of the report for review and discussion by the national council. Patricia Warren, the project manager, provided valuable editorial support. The staff drew on the advice of many of their AAAS colleagues, particularly Audrey Champagne and Shirley Malcom, and of Michael O'Keefe, who was especially helpful in planning Phase II of the project.

None of this Phase I work would have been possible without the generous financial support provided to the AAAS by the Carnegie Corporation of New York and the Andrew W. Mellon Foundation. Alden Dunham, program chair at Carnegie, was one of the first people in the United States to see the need for this study; throughout the more than four years it has taken to transform a vague idea into the reality of this report, no one has been more committed to the project.

Sheila Widnall
Chair, Board of Directors, American Association
    for the Advancement of Science

NATIONAL COUNCIL ON SCIENCE AND TECHNOLOGY EDUCATION

COCHAIRS

William O. Baker Chairman of the Board (retired), AT&T Bell Laboratories

Margaret L. A. MacVicar Professor of Physical Science, Dean for Undergraduate Education, Massachusetts Institute of Technology

MEMBERS

Paula Apsell Executive Director, NOVA-WGBH, Boston

Francisco J. Ayala Distinguished Professor of Biological Sciences, University of California, Irvine

F. Herbert Bormann Oastler Professor of Forest Ecology, School of Forestry and Environmental Studies, Yale University

Margaret Burbidge University Professor, Center for Astrophysics and Space Sciences, University of California, San Diego

Ernestine Friedl James B. Duke Professor of Anthropology, Duke University

Robert Glaser Director of the Learning Research and Development Center, University Professor of Psychology and Education, University of Pittsburgh

Judith Lanier Dean, College of Education, Michigan State University

Arturo Madrid President, Tomas Rivera Center, Claremont Graduate School, Claremont College

Kenneth R. Manning Professor of the History of Science, Massachusetts Institute of Technology

Ray Marshall Professor of Economics and Public Affairs, LBJ School of Public Affairs, University of Texas at Austin

Walter E. Massey Vice President for Research and for Argonne National Laboratory, University of Chicago

Alice Moses Program Director for Instructional Materials Development, Division of Materials, Development, Research and Informal Science Education, National Science Foundation

Frederick Mosteller Professor Emeritus of Mathematical Statistics, Harvard University

Gilbert S. Omenn Dean, School of Public Health and Community Medicine, University of Washington

Gerard Piel Chairman Emeritus, *Scientific American*

George C. Pimentel Professor of Chemistry, University of California, Berkeley

Robert E. Pollack Dean and Professor of Biological Sciences, Columbia College, Columbia University

Henry O. Pollak Former Assistant Vice President, Mathematical Communications and Computer Sciences Research Laboratory, Bell Communications Research

F. James Rutherford Chief Education Officer, American Association for the Advancement of Science

Ted Sanders State Superintendent of Education, Illinois

Albert Shanker President, American Federation of Teachers

Raymond Siever Professor of Geology, Harvard University

Howard Simons Curator of the Nieman Fellowships, Harvard University

Maxine F. Singer President, Carnegie Institution of Washington, and Scientist Emeritus, National Cancer Institute

PROJECT 2061 STAFF

F. James Rutherford Project Director

Andrew Ahlgren Associate Project Director

Patricia S. Warren Project Manager

Carol Holmes Secretary

Gwen McCutcheon Secretary

NOTICE

Six reports have been written as part of the first phase of Project 2061, a long-term, multiphase undertaking of the American Association for the Advancement of Science designed to help reform science, mathematics, and technology education in the United States.

These reports consist of an overview report and five panel reports.

The overview report is *Science for All Americans,* written by the AAAS Project 2061 staff in consultation with the National Council on Science and Technology Education.

The titles of the panel reports are as follows:

• *Biological and Health Sciences: Report of the Project 2061 Phase I Biological and Health Sciences Panel,* by Mary Clark

• *Mathematics: Report of the Project 2061 Phase I Mathematics Panel,* by David Blackwell and Leon Henkin

• *Physical and Information Sciences and Engineering: Report of the Project 2061 Phase I Physical and Information Sciences and Engineering Panel,* by George Bugliarello

• *Social and Behavioral Sciences: Report of the Project 2061 Phase I Social and Behavioral Sciences Panel,* by Mortimer Appley and Winifred B. Maher

• *Technology: Report of the Project 2061 Phase I Technology Panel,* by James R. Johnson

For information on ordering all six reports, please contact Project 2061, the American Association for the Advancement of Science, 1333 H Street NW, Washington, D.C. 20005.

SUMMARY

SUMMARY

Science for All Americans is about scientific literacy. It consists of a set of recommendations by the National Council on Science and Technology Education—a distinguished group of scientists and educators appointed by the American Association for the Advancement of Science—on what understandings and habits of mind are essential for all citizens in a scientifically literate society.

Scientific literacy—which embraces science, mathematics, and technology—has emerged as a central goal of education. Yet the fact is that general scientific literacy eludes us in the United States. A cascade of recent studies has made it abundantly clear that by both national standards and world norms, U.S. education is failing to adequately educate too many students—and hence failing the nation. By all accounts, America has no more urgent priority than the reform of education in science, mathematics, and technology.

But educational reform cannot simply be legislated. It will take time, determination, collaboration, resources, and leadership. It will take daring and experimentation. And it will take a shared national vision of what Americans want their schools to achieve. *Science for All Americans*—part of a AAAS initiative called Project 2061—is intended to help in the formulation of that vision.

In preparing its recommendations, the national council drew on the reports of five independent scientific panels. In addition, the national council sought the advice of a large and diverse array of consultants and reviewers—scientists, engineers, mathematicians, historians, and educators. In all, the process took more than three years, involved hundreds of individuals, and culminated in the unanimous approval of *Science for All Americans* by the board of directors of the AAAS.

PROJECT 2061 AND SCIENTIFIC LITERACY

Project 2061 involves a three-phase plan of purposeful and sustained action that will contribute to the critically needed reform of education in science, mathematics, and technology:

• Phase I focused on the substance of scientific literacy. Its purpose was to establish a conceptual base for reform by spelling out the knowledge, skills, and attitudes all students should acquire as a consequence of their total school experience from kindergarten through high school. *Science for All Americans* and the reports of the scientific panels are the chief products of this phase.

• Phase II involves teams of educators and scientists transforming *Science for All Americans* into several alternative curriculum models for the use of school districts and states. During this phase, the project is also drawing up blueprints for reform related to teacher education, teaching materials and technologies, testing, the organization of schooling, educational policies, and educational research.

• Phase III will be a widespread collaborative effort, lasting a decade or longer, in which many groups active in educational reform will use the resources of Phases I and II to move the nation toward scientific literacy.

RECOMMENDATIONS

A fundamental premise of Project 2061 is that the schools do not need to be asked to teach more and more content, but rather to focus on what is essential to scientific literacy and to teach it more effectively. Accordingly, the national council's recommendations for a common core of learning are limited to the ideas and skills having the greatest scientific and educational significance for scientific literacy.

Science for All Americans is based on the belief that the scientifically literate person is one who is aware that science, mathematics, and technology are interdependent human enterprises with strengths and limitations; understands key concepts and principles of science; is familiar with the natural world and recognizes both its diversity and unity; and uses scientific knowledge and scientific ways of thinking for individual and social purposes.

The national council's specific recommendations constitute the bulk of the report. The first three chapters focus on the nature of science, mathematics, and technology as human enterprises, and on how they resemble and differ from one another. The next six chapters present basic knowledge about the world—the shaping of the physical setting, the evolution and characteristics of life forms, the dynamics of human society, and other key aspects of the world—as seen through the eyes of science and mathematics and as shaped by technology. The following two chapters set forth what people should know about great episodes in the history of the scientific endeavor and about key crosscutting themes—such as systems, patterns of change, and scale—that enable people to understand how the world works. The final chapter of the content recommendations focuses on the habits of mind that are essential for scientific literacy.

The council's recommendations cover a broad array of topics. Many of these topics are already common in school curricula (for example, the structure of matter, the basic functions of cells, prevention of disease, communications technology, and different uses of numbers). However, the treatment of such topics tends to differ from the traditional in two ways.

One difference is that boundaries between traditional subject-matter categories are softened and connections are emphasized. Transformations of energy, for example, occur in physical, biological, and technological systems, and evolutionary change appears in stars, organisms, and societies.

A second difference is that the amount of detail that students are expected to retain is considerably less than in traditional science, mathematics, and technology courses. Ideas and thinking skills are emphasized at the expense of specialized vocabulary and memorized procedures. Sets of ideas are chosen that not only make some satisfying sense at a simple level but also provide a lasting foundation for learning more. Details are treated as a means of enhancing, not guaranteeing, understanding of a general idea. The council believes, for example, that basic scientific literacy implies knowing that the chief function of living cells is assembling protein molecules according to instructions coded in DNA molecules, but does not imply knowing the terms "ribosome" or "deoxyribonucleic acid," or knowing what messenger RNA is and how it relates to DNA.

The national council's recommendations include some topics that are not common in school curricula. Among these topics are the nature of the scientific enterprise, including how science, mathematics, and technology relate to one another and to the social system in general. The council also calls for some knowledge of the most important episodes in the history of science and technology, and of the major conceptual themes that run through almost all scientific thinking.

BRIDGES TO THE FUTURE

Certain next steps are essential if the nation is going to make significant headway toward realizing the goals expressed in *Science for All Americans*. Those steps should reflect the following considerations:

• To ensure the scientific literacy of all students, curricula must be changed to reduce the sheer amount of material covered; to weaken or eliminate rigid subject-matter boundaries; to pay more attention to the connections among science, mathematics, and technology; to present the scientific endeavor as a social enterprise that strongly influences—and is influenced by—human thought and action; and to foster scientific ways of thinking.

• The effective teaching of science, mathematics, and technology (or any other body of knowledge and skills) must be based on learning principles that derive from systematic research and from well-tested craft experience. Moreover, teaching related to scientific literacy needs to be consistent with the spirit and character of scientific inquiry and with scientific values. This suggests such approaches as starting with questions about phenomena rather than with answers to be learned; engaging students actively in the use of hypotheses, the collection and use of evidence, and the design of investigations and processes; and placing a premium on students' curiosity and creativity.

• Educational reform must be comprehensive, focusing on the learning needs of all children, covering all grades and subjects, and dealing with all components and aspects of the educational system. It will also require that positive conditions for change be established and that public support for reform be sustained for a decade or longer.

• Reform must be collaborative. It must involve administrators, university faculty members, and community, business, labor, and political leaders, as well as teachers, parents, and students themselves.

In support of collaborative reform, *Science for All Americans* concludes with an agenda for action that suggests steps individuals, institutions, organizations, and government agencies can take to work together toward reform. For its part, Project 2061 will proceed with Phases II and III and also will continue to do what it can to keep scientific literacy and educational reform on the agenda of educators, scientists, policymakers, and the public.

There are no valid reasons—intellectual, social, or economic—why the United States cannot transform its schools to make scientific literacy possible for all students. What is required is national commitment, determination, and a willingness to work together toward common goals.

PART I:

EDUCATION
FOR A CHANGING
FUTURE

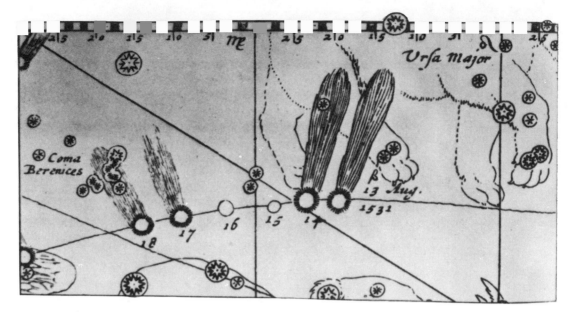

Peter Apian's observations of Comet Halley in 1531.

PART I
INTRODUCTION

Scientific literacy—which embraces literacy in science, mathematics, and technology—has emerged as a central goal of education. Yet the fact is that scientific literacy eludes us in the United States. A cascade of recent studies (see Appendix B) has made it abundantly clear that by both national standards and international norms, U.S. education is failing to adequately educate too many students—and hence failing the nation. The reform of science, mathematics, and technology education must rank as one of America's highest priorities.

Scientific societies have responded to this challenge in many ways. As one such response, the American Association for the Advancement of Science has initiated Project 2061, a long-range, multiphase effort designed to help the nation achieve scientific literacy. It was started in 1985, a year when Comet Halley happened to be in the earth's vicinity. That coincidence prompted the project's name, for it was realized that the children who would live to see the return of the comet in 2061 would soon be starting their school years.

Project 2061 is based on these convictions:

• All children need and deserve a basic education in science, mathematics, and technology that prepares them to live interesting and productive lives.

• World norms for what constitutes a basic education have changed radically in response to the rapid growth of scientific knowledge and technological power.

• U.S. schools have yet to act decisively enough in preparing young people—especially minority children, on whom the future of America is coming to depend—for a world shaped by science and technology.

• Sweeping changes in the entire educational system from kindergarten through twelfth grade will have to be made if the United States is to become a nation of scientifically literate citizens.

• A necessary first step in achieving systematic reform in science, mathematics, and technology education is reaching a clear understanding of what constitutes scientific literacy.

This view led Project 2061 to focus its initial effort on the need to establish learning goals in science, mathematics, and technology for all American children. A distinguished group of scientists and educators—the National Council on Science and Technology Education—was appointed by the AAAS board of directors specifically to address the question of what understandings and habits of mind are essential for all citizens in a scientifically literate society. The council's response is to be found in the recommendations of this report.

In preparing its recommendations, the council drew on the reports of five independent scientific panels. In addition it had the benefit of the advice of a large and diverse array of consultants and reviewers—scientists, engineers, mathematicians, historians, and educators. (See Appendix A for the names of the panelists and other participants in Phase I of Project 2061.) In all, the process took more than three years, involved hundreds of individuals, and culminated in the unanimous approval of *Science for All Americans* by the AAAS board of directors.

In Part III of this report, there is a discussion of other matters that bear on or follow from the national council's recommendations. These include comments on principles of teaching that are consistent with the council's recommendations, on what it will take to achieve reform on a national scale, and on ways in which the recommendations in *Science for All Americans* can be used by individuals, institutions, and organizations in their efforts to help transform the United States into a scientifically literate nation.

THE NEED FOR SCIENTIFIC LITERACY

Education has no higher purpose than preparing people to lead personally fulfilling and responsible lives. For its part, science education—meaning education in science, mathematics, and technology—should help students to develop the understandings and habits of mind they need to become compassionate human beings able to think for themselves and to face life head on. It should equip them also to participate thoughtfully with fellow citizens in building and protecting a society that is open, decent, and vital. America's future—its ability to create a truly just society, to sustain its economic vitality, and to remain secure in a world torn by hostilities—depends more than ever on the character and quality of the education that the nation provides for all of its children.

There is more at stake, however, than individual self-fulfillment and the immediate national interest of the United States. The most serious problems that humans now face are global: unchecked population growth in many parts of the world, acid rain, the shrinking of tropical rain forests and other great sources of species diversity, the pollution of the environment, disease, social strife, the extreme inequities in the distribution of the earth's wealth, the huge investment of human intellect and scarce resources in preparing for and conducting war, the ominous shadow of nuclear holocaust—the list is long, and it is alarming.

What the future holds in store for individual human beings, the nation, and the world depends largely on the wisdom with which humans use science and technology. But that, in turn, depends on the character, distribution, and effectiveness of the education that people receive. Briefly put, the national council's argument is this:

• Science, energetically pursued, can provide humanity with the knowledge of the biophysical environment and of social behavior that it needs to develop effective solutions to its global and local problems; without that knowledge, progress toward a safe world will be unnecessarily handicapped.

• By emphasizing and explaining the dependency of living things on each other and on the physical environment, science fosters the kind of intelligent respect for nature that should inform decisions on the uses of technology; without that respect, we are in danger of recklessly destroying our life-support system.

- Scientific habits of mind can help people in every walk of life to deal sensibly with problems that often involve evidence, quantitative considerations, logical arguments, and uncertainty; without the ability to think critically and independently, citizens are easy prey to dogmatists, flimflam artists, and purveyors of simple solutions to complex problems.

- Technological principles relating to such topics as the nature of systems, the importance of feedback and control, the cost-benefit-risk relationship, and the inevitability of side effects give people a sound basis for assessing the use of new technologies and their implications for the environment and culture; without an understanding of those principles, people are unlikely to move beyond consideration of their own immediate self-interest.

- Although many pressing global and local problems have technological origins, technology provides the tools for dealing with such problems, and the instruments for generating, through science, crucial new knowledge; without the continuous development and creative use of new technologies, society will limit its capacity for survival and for working toward a world in which the human species is at peace with itself and its environment.

- The life-enhancing potential of science and technology cannot be realized unless the public in general comes to understand science, mathematics, and technology and to acquire scientific habits of mind; without a scientifically literate population, the outlook for a better world is not promising.

THE CURRENT SITUATION

Most Americans are not scientifically literate. One only has to look at the international studies of educational performance to see that U.S. students rank near the bottom in science and mathematics—hardly what one would expect if the schools were doing their job well. The most recent international mathematics study has reported, for instance, that U.S. students are well below the international level in problem solving, and the latest study of National Assessment of Educational Progress has found that despite some small recent gains, the average performance of 17-year-olds in 1986 remained substantially lower than it had been in 1969.

The United States should be able to do better. It is, after all, a prosperous nation that claims to value public education as the foundation of democracy. And it has deliberately staked its future well-being on its competence—even leadership—in science and technology. Surely it is reasonable, therefore, to expect this commitment to show up in the form of a modern, well-supported school system staffed by highly qualified teachers and administrators. And surely the curriculum in such schools should feature science, mathematics, and technology for all students. In fact, however, the situation existing in far too many states and school districts is quite different:

- Few elementary school teachers have even a rudimentary education in science and mathematics, and many junior and senior high school teachers of science and mathematics do not meet reasonable standards of preparation in those fields. Unfortunately, such deficiencies have long been tolerated by the institutions that prepare teachers, the public bodies that license them, the schools that hire them and give them their assignments, and even the teaching profession itself.

- Teachers of science and mathematics have crushing teaching loads that make it nearly impossible for them to perform well, no matter how excellent their preparation may have been. This burden is made worse by the almost complete absence of a modern support system to back them up. As the world approaches the twenty-first century, the schools of America—when it comes to the deployment of people, time, and technology—seem to be still stuck in the nineteenth century.

- The present science textbooks and methods of instruction, far from helping, often actually impede progress toward scientific literacy. They emphasize the learning of answers more than the exploration of questions, memory at the expense of critical thought, bits and pieces of information instead of understandings in context, recitation over argument, reading in lieu of doing. They fail to encourage students to work together, to share ideas and information freely with each other, or to use modern instruments to extend their intellectual capabilities.

- The present curricula in science and mathematics are overstuffed and undernourished. Over the decades, they have grown with little restraint, thereby overwhelming teachers and students and making it difficult for them to keep track of what science, mathematics, and technology is truly essential. Some topics are taught over and over again in needless detail; some that are of equal or greater importance to scientific literacy—often from the physical and social sciences and from technology—are absent from the curriculum or are reserved for only a few students.

To turn this situation around will take determination, resources, leadership, and time. The world has changed in such a way that scientific literacy has become necessary for everyone, not just a privileged few; science education will have to change to make that possible. We are all responsible for the current deplorable state of affairs in education, and it will take all of us to reform it. Project 2061 hopes to contribute to that national effort.

THE THREE PHASES OF PROJECT 2061

Because the work of Project 2061 is expected to span a decade or more, it has been organized into three phases.

Phase I of the project has attempted to establish a conceptual base for reform by defining the knowledge, skills, and attitudes all students should acquire as a consequence of their total school experience, from kindergarten through high school. Drawing on ideas proposed by panels of prestigious scientists, mathematicians, and engineers, this report, *Science for All Americans,* is the culmination of that effort.

During Phase II of Project 2061, now under way, teams of educators and scientists are transforming this report into blueprints for action. The main purpose of the second phase of the project is to produce a variety of curriculum models that school districts and states can use as they undertake to reform the teaching of science, mathematics, and technology. Phase II will also specify the characteristics of other reforms needed to make it possible for new curricula to work: teacher education, testing policies and practices, new materials and modern technologies, the organization of schooling, state and local policies, and research. (See Chapter 15, "Next Steps," for a more detailed account of Phase II.)

In Phase III, the project will collaborate with scientific societies, educational organizations and institutions, and other groups involved in the reform of science, mathematics, and technology education in a nationwide effort to turn the Phase II blueprints into educational practice.

PART II:

RECOMMENDATIONS OF THE NATIONAL COUNCIL

Comet Halley in the desert sky, 1986.

INTRODUCTION

The National Council on Science and Technology Education was asked to answer this question: Out of all the possibilities, what knowledge, skills, and habits of mind associated with science, mathematics, and technology should all Americans have by the time they leave school? As the council quickly found out, this deceptively easy question had no easy or self-evident answer. The task was made even more difficult by the constraints placed on the council:

• Consider the question from base zero—nothing should be automatically included in the recommendations, no matter how long it may have been imbedded in curricula, textbooks, and exams.

• Consider possibilities across all of science, mathematics, and technology, but do not strive necessarily for equal portions of each.

• Come up with learning goals that are modest enough to make sense for all students (including those who do not ordinarily perform well academically) but that are nevertheless ambitious enough to raise the sights of students and teachers.

• Study the reports of the scientific panels carefully and take into account other viewpoints as well, but then reach independent conclusions rather than seek a compromise among all possible views.

Subsequently, the council members did reach a strong consensus. Although each member might have wished to add or subtract a bit or to change the order or language somewhat (as many of the consultants and reviewers may have), such differences were generally peripheral. The council believes that taken as a whole, the recommendations that follow provide a forward-looking answer to the question posed by Project 2061 and constitute an answer that fairly reflects the views of the scientific community.

The recommendations are presented in 12 chapters that thematically cover four major categories:

• Chapters 1 through 3 deal with the nature of science, mathematics, and technology—collectively, the scientific endeavor—as human enterprises.

• Chapters 4 through 9 cover basic knowledge about the world as currently seen from the perspective of science and mathematics and as shaped by technology.

• Chapters 10 and 11 present what people should understand about some of the great episodes in the history of the scientific endeavor and about some crosscutting themes that can serve as tools for thinking about how the world works.

• Chapter 12 lays out the habits of mind that are essential for scientific literacy.

In considering these recommendations, it is important to keep in mind some of the special features of the report:

The Recommendations Reflect a Broad Definition of Scientific Literacy

Scientific literacy—which encompasses mathematics and technology as well as the natural and social sciences—has many facets. These include being familiar with the natural world and respecting its unity; being aware of some of the important ways in which mathematics, technology, and the sciences depend upon one another; understanding some of the key concepts and principles of science; having a capacity for scientific ways of thinking; knowing that science, mathematics, and technology are human enterprises, and knowing what that implies about their strengths and limitations; and being able to use scientific knowledge and ways of thinking for personal and social purposes.

Some of these facets of scientific literacy are addressed only in specific places in the report, whereas others are woven into the text of the chapters. It is essential, therefore, that the recommendations be viewed in their entirety as a multifaceted discussion of scientific literacy.

The Recommendations in This Report Apply to All Students

The set of recommendations constitutes a common core of learning in science, mathematics, and technology for all young people, regardless of their social circumstances and career aspirations. In particular, the recommendations pertain to those who in the past have largely been bypassed in science and mathematics education: ethnic and language minorities and girls. The recommendations do not include every interesting topic that was suggested and do not derive from diluting the traditional college preparatory curriculum. Nevertheless, the recommendations are deliberately ambitious, for it would be worse to underestimate what students can learn than to expect too much. The national council is convinced that—given clear goals, the right resources, and good teaching throughout 13 years of school—essentially all students (operationally meaning 90 percent or more) will be able to reach all of the recommended learning goals (meaning at least 90 percent) by the time they graduate from high school.

At the same time, however, no student should be limited to the common core of learning spelled out in this report. In response to special interests and skills, some students will want to gain a more sophisticated understanding of the topics than what is suggested here, and some will want to pursue topics not included here at all. A well-designed curriculum will be able to serve those special needs without sacrificing a commitment to a common core of learning in science, mathematics, and technology.

The Recommendations Have Been Selected on the Basis of Both Scientific and Human Significance

The schools do not need to be asked to teach more and more content, but to teach less in order to teach it better. By concentrating on fewer topics, teachers can introduce ideas gradually, in a variety of contexts, reinforcing and extending them as students mature.

Students will end up with richer insights and deeper understandings than they could hope to gain from a superficial exposure to more topics than they can assimilate. The problem for curriculum developers, therefore, is much less what to add than what to eliminate.

Reversing the accretion of material over scores of years is thus a major goal of Project 2061. But addressing this goal has meant making choices. The criteria for identifying a common core of learning in science, mathematics, and technology were both scientific and educational. Consideration was given first to the ideas that seemed to be of unusual scientific importance, because there is simply too much knowledge for anyone to acquire in a lifetime, let alone 13 years. This meant favoring content that has had great influence on what is worth knowing now and what will still be worth knowing decades hence, and ruling out topics mainly of only passing technical interest or limited scientific scope. In particular, concepts were chosen that could serve as a lasting foundation on which to build more knowledge over a lifetime. The choices then had to meet important criteria having to do with human life and with the broad goals that justify universal public education in a free society. The criteria were:

Utility. Will the proposed content—knowledge or skills—significantly enhance the graduate's long-term employment prospects? Will it be useful in making personal decisions?

Social Responsibility. Is the proposed content likely to help citizens participate intelligently in making social and political decisions on matters involving science and technology?

The Intrinsic Value of Knowledge. Does the proposed content present aspects of science, mathematics, and technology that are so important in human history or so pervasive in our culture that a general education would be incomplete without them?

Philosophical Value. Does the proposed content contribute to the ability of people to ponder the enduring questions of human meaning such as life and death, perception and reality, the individual good versus the collective welfare, certainty and doubt?

Childhood Enrichment. Will the proposed content enhance childhood (a time of life that is important in its own right and not solely for what it may lead to in later life)?

The Recommendations Are Neither All New Nor Intended to Be Fixed for All Time

In formulating recommendations, no attempt was made to either seek novelty or avoid it. The task was to identify a minimal core of critical understandings and skills, whether or not they happen to be part of current school curricula. The recommendations do not constitute the only possible ones, and indeed there were differences among the participants in this project on various topics. The council does believe, however, that the recommendations make good sense and that they offer a sound basis for designing curricula in science, mathematics, and technology.

But science, mathematics, and technology are continually in flux—holding onto some ideas and ways of doing things, reshaping or discarding some, adding others. The time will inevitably come—sooner in some areas than others—when the recommendations will need to be revised to accommodate new knowledge. Furthermore, as educators and scientists work together in Phase II of Project 2061 to design curriculum models based on this report, they are likely to reach their own conclusions on the appropriateness of these recommendations and to suggest changes. In any case, the recommendations are not presented to set up a new and unalterable orthodoxy, but rather to provide a credible resource for the Phase II development, to provoke lively debate on the question of the content of education, and to catalyze curriculum reform.

This Report Is Not a Curriculum Document or a Textbook

The reader should not expect to find recommendations in this report on what should be taught in any particular course or at any grade level. The report deals only with learning goals—what students should remember, understand, and be able to do after they have left school as a residue of their total school experience—and not with how to organize the curriculum to achieve them. Neither is the presentation of recommendations meant to instruct the reader as a text does. No linear presentation of topics can satisfactorily represent the connectedness of ideas and experiences that would be essential in an actual curriculum or textbook.

The Recommendations Are Intended to Convey the Levels of Understanding Appropriate for All People

For most educational purposes, broad generalizations (such as "everyone should know how science and technology are related") are no more useful than are long lists of specific topics (atoms, cells, planets, graphs, etc.). Neither approach reveals what is to be learned, and both require the reader to guess what level of sophistication is intended. Thus, the specific recommendations in this report are framed in enough detail to convey the levels of understanding and the contexts of understanding intended. The recommendations have been formulated under four levels of generalization:

Chapters. Each chapter deals with a major set of related topics. Collectively, the chapter titles lay out a conceptual framework for understanding science that people can use throughout their lives as they gain new knowledge about the world.

Headings. Within each chapter, headings such as "Forces That Shape the Earth" or "Interdependence of Life" identify the conceptual categories that all students should be familiar with. A list of all the headings would provide an approximate answer to the question of the scope, but not the content, of the specific recommendations.

Paragraphs. Under each heading are paragraphs that express the residual knowledge, insights, and skills that people should possess after the details have faded from memory. If high school graduates were interviewed about a topic—"Information Processing," say—they should be able to come up, in their own words, with the ideas sketched in the paragraphs under that heading.

Vocabulary. The language of the recommendations is intended to convey the level of learning advocated. The recommendations are written for today's educated adults, not students—but the technical vocabulary is limited to what would be desirable for all students to command, as a minimum, by the time they finish school. This vocabulary should be viewed as a product of a sound education in science, mathematics, and technology, but not its main purpose.

In sum, the recommendations are—to different degrees of specificity—implicit in the titles, headings, text, and vocabulary of the 12 chapters that follow. Yet there is no way, in so short a document, to convey the quality of knowledge envisaged across the full range of topics. This quality—the way in which something is known—depends largely on how it is learned. In this regard, the discussion of learning and teaching in Part III provides a perspective for understanding the nature of the recommendations themselves.

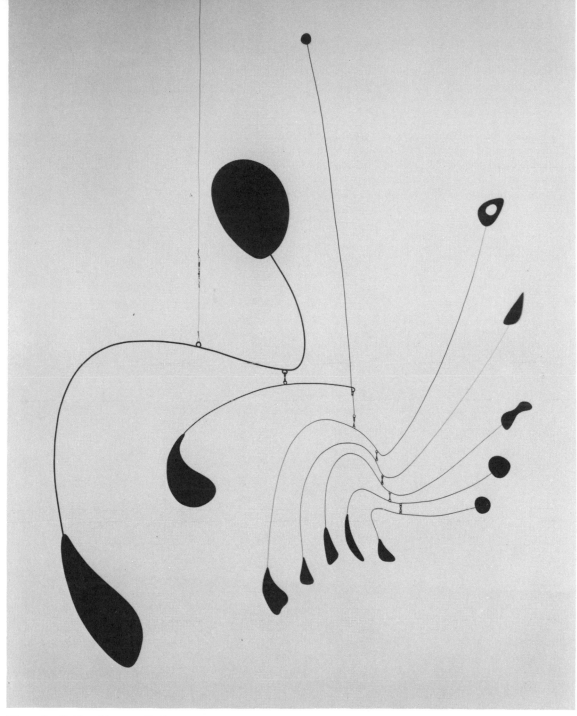

Alexander Calder, *Hanging Spider* (1940).

CHAPTER 1

THE NATURE OF SCIENCE

Over the course of human history, people have developed many interconnected and validated ideas about the physical, biological, psychological, and social worlds. Those ideas have enabled successive generations to achieve an increasingly comprehensive and reliable understanding of the human species and its environment. The means used to develop these ideas are particular ways of observing, thinking, experimenting, and validating. These ways represent a fundamental aspect of the nature of science and reflect how science tends to differ from other modes of knowing.

It is the union of science, mathematics, and technology that forms the scientific endeavor and that makes it so successful. Although each of these human enterprises has a character and history of its own, each is dependent on and reinforces the others. Accordingly, the first three chapters of recommendations draw portraits of science, mathematics, and technology that emphasize their roles in the scientific endeavor and reveal some of the similarities and connections among them.

This chapter lays out recommendations for what knowledge of the way science works is requisite for scientific literacy. The chapter focuses on three principal subjects: the scientific world view, scientific methods of inquiry, and the nature of the scientific enterprise. Chapters 4 through 9 present views of the world as depicted by current science; Chapter 10, "Historical Perspectives," covers key episodes in the development of science; and Chapter 11, "Common Themes," pulls together ideas that cut across all these views of the world.

RECOMMENDATIONS

THE SCIENTIFIC WORLD VIEW

Scientists share certain basic beliefs and attitudes about what they do and how they view their work. These have to do with the nature of the world and what can be learned about it.

The World Is Understandable

Science presumes that the things and events in the universe occur in consistent patterns that are comprehensible through careful, systematic study. Scientists believe that through the use of the intellect, and with the aid of instruments that extend the senses, people can discover patterns in all of nature.

Science also assumes that the universe is, as its name implies, a vast single system in which the basic rules are everywhere the same. Knowledge gained from studying one part of the universe is applicable to other parts. For instance, the same principles of motion and gravitation that explain the motion of falling objects on the surface of the earth also explain the motion of the moon and the planets. With some modifications over the years, the same principles of motion have applied to other forces—and to the motion of everything, from the smallest nuclear particles to the most massive stars, from sailboats to space vehicles, from bullets to light rays.

Scientific Ideas Are Subject to Change

Science is a process for producing knowledge. The process depends both on making careful observations of phenomena and on inventing theories for making sense out of those observations. Change in knowledge is inevitable because new observations may challenge prevailing theories. No matter how well one theory explains a set of observations, it is possible that another theory may fit just as well or better, or may fit a still wider range of observations. In science, the testing and improving and occasional discarding of theories, whether new or old, goes on all the time. Scientists assume that even if there is no way to secure complete and absolute truth, increasingly accurate approximations can be made to account for the world and how it works.

Scientific Knowledge Is Durable

Although scientists reject the notion of attaining absolute truth and accept some uncertainty as part of nature, most scientific knowledge is durable. The modification of ideas, rather than their outright rejection, is the norm in science, as powerful constructs tend to survive and grow more precise and to become widely accepted. For example, in formulating the theory of relativity, Albert Einstein did not discard the Newtonian laws of motion but rather showed them to be only an approximation of limited application within a more general concept. (The National Aeronautics and Space Administration uses Newtonian mechanics, for instance, in calculating satellite trajectories.) Moreover, the growing ability of scientists to make accurate predictions about natural phenomena provides convincing evidence that we really are gaining in our understanding of how the world works. Continuity and stability are as characteristic of science as change is, and confidence is as prevalent as tentativeness.

Science Cannot Provide Complete Answers to All Questions

There are many matters that cannot usefully be examined in a scientific way. There are, for instance, beliefs that—by their very nature—cannot be proved or disproved (such as the existence of super-natural powers and beings, or the true purposes of life). In other cases, a scientific approach that may be valid is likely to be rejected as irrelevant by people who hold to certain beliefs (such as in miracles, fortune-telling, astrology, and superstition). Nor do scientists have the means to settle issues concerning good and evil, although they can sometimes contribute to the discussion of such issues by identifying the likely consequences of particular actions, which may be helpful in weighing alternatives.

SCIENTIFIC INQUIRY

Fundamentally, the various scientific disciplines are alike in their reliance on evidence, the use of hypothesis and theories, the kinds of logic used, and much more. Nevertheless, scientists differ greatly from one another in what phenomena they investigate and in how they go about their work; in the reliance they place on historical data or on experimental findings and on qualitative or quantitative methods; in their recourse to fundamental principles; and in how much they draw on the findings of other sciences. Still, the exchange of techniques, information, and concepts goes on all the time among scientists, and there are common understandings among them about what constitutes an investigation that is scientifically valid.

Scientific inquiry is not easily described apart from the context of particular investigations. There simply is no fixed set of steps that scientists always follow, no one path that leads them unerringly to scientific knowledge. There are, however, certain features of science that give it a distinctive character as a mode of inquiry. Although those features are especially characteristic of the work of professional scientists, everyone can exercise them in thinking scientifically about many matters of interest in everyday life.

Science Demands Evidence

Sooner or later, the validity of scientific claims is settled by referring to observations of phenomena. Hence, scientists concentrate on getting accurate data. Such evidence is obtained by observations and measurements taken in situations that range from natural settings (such as a forest) to completely contrived ones (such as the laboratory). To make their observations, scientists use their own senses, instruments (such as microscopes) that enhance those senses, and instruments that tap characteristics quite different from what humans can sense (such as magnetic fields). Scientists observe passively (earthquakes, bird migrations), make collections (rocks, shells), and actively probe the world (as by boring into the earth's crust or administering experimental medicines).

In some circumstances, scientists can control conditions deliberately and precisely to obtain their evidence. They may, for example, control the temperature, change the concentration of chemicals, or choose which organisms mate with which others. By varying just one condition at a time, they can hope to identify its exclusive effects on what happens, uncomplicated by changes in other conditions. Often, however, control of conditions may be impractical (as in studying stars), or unethical (as in studying people), or likely to distort the natural phenomena (as in studying wild animals in captivity). In such cases, observations have to be made over a sufficiently wide range of naturally occurring conditions to infer what the influence of various factors might be. Because of this reliance on evidence, great value is placed on the development of better instruments and techniques of observation, and the findings of any one investigator or group are usually checked by others.

Science Is a Blend of Logic and Imagination

Although all sorts of imagination and thought may be used in coming up with hypotheses and theories, sooner or later scientific arguments must conform to the principles of logical reasoning—that is, to testing the validity of arguments by applying certain criteria of inference, demonstration, and common sense. Scientists may often disagree about the value of a particular piece of evidence, or about the appropriateness of particular assumptions that are made—and therefore disagree about what conclusions are justified. But they tend to agree about the principles of logical reasoning that connect evidence and assumptions with conclusions.

Scientists do not work only with data and well-developed theories. Often, they have only tentative hypotheses about the way things may be. Such hypotheses are widely used in science for choosing what data to pay attention to and what additional data to seek, and for guiding the interpretation of data. In fact, the process of formulating and testing hypotheses is one of the core activities of scientists. To be useful, a hypothesis should suggest what evidence would support it and what evidence would refute it. A hypothesis that cannot in principle be put to the test of evidence may be interesting, but it is not scientifically useful.

The use of logic and the close examination of evidence are necessary but not usually sufficient for the advancement of science. Scientific concepts do not emerge automatically from data or from any amount of analysis alone. Inventing hypotheses or theories to imagine how the world works and then figuring out how they can be put to the test of reality is as creative as writing poetry, composing music, or designing skyscrapers. Sometimes discoveries in science are made unexpectedly, even by accident. But knowledge and creative insight are usually required to recognize the meaning of the unexpected. Aspects of data that have been ignored by one scientist may lead to new discoveries by another.

Science Explains and Predicts

Scientists strive to make sense of observations of phenomena by inventing explanations for them that use, or are consistent with, currently accepted scientific principles. Such explanations—theories—may be either sweeping or restricted, but they must be logically sound and incorporate a significant body of scientifically valid observations. The credibility of scientific theories often comes from their ability to show relationships among phenomena that previously seemed unrelated. The theory of moving continents, for example, has grown in credibility as it has shown relationships among such diverse phenomena as earthquakes, volcanoes, the match between types of fossils on different continents, the shapes of continents, and the contours of the ocean floors.

The essence of science is validation by observation. But it is not enough for scientific theories to fit only the observations that are already known. Theories should also fit additional observations that were not used in formulating the theories in the first place; that is, theories should have predictive power. Demonstrating the predictive power of a theory does not necessarily require the prediction of events in the future. The predictions may be about evidence from the past that has not yet been found or studied. A theory about the origins of human beings, for example, can be tested by new discoveries of human-like fossil remains. This approach is clearly necessary for reconstructing the events in the history of the earth or of the life forms on it. It is also necessary for the study of processes that usually occur very slowly, such as the building of mountains or the aging of stars. Stars, for example, evolve more slowly than we can usually observe. Theories of the evolution of stars, however, may predict unsuspected relationships between features of starlight that can then be sought in existing collections of data about stars.

Scientists Try to Identify and Avoid Bias

When faced with a claim that something is true, scientists respond by asking what evidence supports it. But scientific evidence can be biased in how the data are interpreted, in the recording or reporting of the data, or even in the choice of what data to consider in the first place. Scientists' nationality, sex, ethnic origin, age, political convictions, and so on may incline them to look for or emphasize one or another kind of evidence or interpretation. For example, for many years the study of primates—by male scientists—focused on the competitive social behavior of males. Not until female scientists entered the field was the importance of female primates' community-building behavior recognized.

Bias attributable to the investigator, the sample, the method, or the instrument may not be completely avoidable in every instance, but scientists want to know the possible sources of bias and how bias is likely to influence evidence. Scientists want, and are expected, to be as alert to possible bias in their own work as in that of other scientists, although such objectivity is not always

achieved. One safeguard against undetected bias in an area of study is to have many different investigators or groups of investigators working in it.

Science Is Not Authoritarian

It is appropriate in science, as elsewhere, to turn to knowledgeable sources of information and opinion, usually people who specialize in relevant disciplines. But esteemed authorities have been wrong many times in the history of science. In the long run, no scientist, however famous or highly placed, is empowered to decide for other scientists what is true, for none are believed by other scientists to have special access to the truth. There are no preestablished conclusions that scientists must reach on the basis of their investigations.

In the short run, new ideas that do not mesh well with mainstream ideas may encounter vigorous criticism, and scientists investigating such ideas may have difficulty obtaining support for their research. Indeed, challenges to new ideas are the legitimate business of science in building valid knowledge. Even the most prestigious scientists have occasionally refused to accept new theories despite there being enough accumulated evidence to convince others. In the long run, however, theories are judged by their results: When someone comes up with a new or improved version that explains more phenomena or answers more important questions than the previous version, the new one eventually takes its place.

THE SCIENTIFIC ENTERPRISE

Science as an enterprise has individual, social, and institutional dimensions. Scientific activity is one of the main features of the contemporary world and, perhaps more than any other, distinguishes our times from earlier centuries.

Science Is a Complex Social Activity

Scientific work involves many individuals doing many different kinds of work and goes on to some degree in all nations of the world. Men and women of all ethnic and national backgrounds participate in science and its applications. These people—scientists and engineers, mathematicians, physicians, technicians, computer programmers, librarians, and others—may focus on scientific knowledge either for its own sake or for a particular practical purpose, and they may be concerned with data gathering, theory building, instrument building, or communicating.

As a social activity, science inevitably reflects social values and viewpoints. The history of economic theory, for example, has paralleled the development of ideas of social justice—at one time, economists considered the optimum wage for workers to be no more than what would just barely allow the workers to survive. Before the twentieth century, and well into it, women and blacks were essentially excluded from most of science by restrictions on their education and employment opportunities; the remarkable few who overcame those obstacles were even then likely to have their work belittled by the science establishment.

The direction of scientific research is affected by informal influences within the culture of science itself, such as prevailing opinion on what questions are most interesting or what methods of investigation are most likely to be fruitful. Elaborate processes involving scientists themselves have been developed to decide which research proposals receive funding, and committees of scientists regularly review progress in various disciplines to recommend general priorities for funding.

Science goes on in many different settings. Scientists are employed by universities, hospitals, business and industry, government, independent research organizations, and scientific associations. They may work alone, in small groups, or as members of large research teams. Their places of work include classrooms, offices, laboratories, and natural field settings from space to the bottom of the sea.

Because of the social nature of science, the dissemination of scientific information is crucial to its progress. Some scientists present their findings and theories in papers that are delivered at meetings or published in scientific journals. Those papers enable scientists to inform others about their work, to expose their ideas to criticism by other scientists, and, of course, to stay abreast of scientific developments around the world.

The advancement of information science (knowledge of the nature of information and its manipulation) and the development of information technologies (especially computer systems) affect all sciences. Those technologies speed up data collection, compilation, and analysis; make new kinds of analysis practical; and shorten the time between discovery and application.

Science Is Organized Into Content Disciplines and Is Conducted in Various Institutions

Organizationally, science can be thought of as the collection of all of the different scientific fields, or content disciplines. From anthropology through zoology, there are dozens of such disciplines. They differ from one another in many ways, including history, phenomena studied, techniques and language used, and kinds of outcomes desired. With respect to purpose and philosophy, however, all are equally scientific and together make up the same scientific endeavor. The advantage of having disciplines is that they provide a conceptual structure for organizing research and research findings. The disadvantage is that their divisions do not necessarily match the way the world works, and they can make communication difficult. In any case, scientific disciplines do not have fixed borders. Physics shades into chemistry, astronomy, and geology, as does chemistry into biology and psychology, and so on. New scientific disciplines (astrophysics and sociobiology, for instance) are continually being formed at the boundaries of others. Some disciplines grow and break into subdisciplines, which then become disciplines in their own right.

Universities, industry, and government are also part of the structure of the scientific endeavor. University research usually emphasizes knowledge for its own sake, although much of it is also directed toward practical problems. Universities, of course, are also particularly committed to educating successive generations of scientists, mathematicians, and engineers. Industries and businesses usually emphasize research directed to practical ends, but many also sponsor research that has no immediately obvious applications, partly on the premise that it will be applied fruitfully in the long run. The federal government funds much of the research in universities and in industry but also supports and conducts research in its many national laboratories and research centers. Private foundations, public-interest groups, and state governments also support research.

Funding agencies influence the direction of science by virtue of the decisions they make on which research to support. Other deliberate controls on science result from federal (and sometimes local) government regulations on research practices that are deemed to be dangerous and on the treatment of the human and animal subjects used in experiments.

There Are Generally Accepted Ethical Principles in the Conduct of Science

Most scientists conduct themselves according to the ethical norms of science. The strongly held traditions of accurate recordkeeping, openness, and replication, buttressed by the critical review of one's work by peers, serve to keep the vast majority of scientists well within the bounds of ethical professional behavior. Sometimes, however, the pressure to get credit for being the first to publish an idea or observation leads some scientists to withhold information or even to falsify their findings. Such a violation of the very nature of science impedes science. When discovered, it is strongly condemned by the scientific community and the agencies that fund research.

Another domain of scientific ethics relates to possible harm that could result from scientific experiments. One aspect is the treatment of live experimental subjects. Modern scientific ethics require that due regard must be given to the health, comfort, and well-being of animal subjects. Moreover, research involving human subjects may be conducted only with the informed consent of the subjects, even if this constraint limits some kinds of potentially important research or influences the results. Informed consent entails full disclosure of the risks and intended benefits of the research and the right to refuse to participate. In addition, scientists must not knowingly subject coworkers, students, the neighborhood, or the community to health or property risks without their knowledge and consent.

The ethics of science also relates to the possible harmful effects of applying the results of research. The long-term effects of science may be unpredictable, but some idea of what applications are expected from scientific work can be ascertained by knowing who is interested in funding it. If, for example, the Department of Defense offers contracts for working on a line of theoretical mathematics, mathematicians may infer that it has application to new military technology and therefore would likely be subject to secrecy measures. Military or industrial secrecy is acceptable to some scientists but not to others. Whether a scientist chooses to work on research of great potential risk to humanity, such as nuclear weapons or germ warfare, is considered by many scientists to be a matter of personal ethics, not one of professional ethics.

Scientists Participate in Public Affairs Both as Specialists and as Citizens

Scientists can bring information, insights, and analytical skills to bear on matters of public concern. Often they can help the public and its representatives to understand the likely causes of events (such as natural and technological disasters) and to estimate the possible effects of projected policies (such as ecological effects of various farming methods). Often they can testify to what is not possible. In playing this advisory role, scientists are expected to be especially careful in trying to distinguish fact from interpretation, and research findings from speculation and opinion; that is, they are expected to make full use of the principles of scientific inquiry.

Even so, scientists can seldom bring definitive answers to matters of public debate. Some issues are too complex to fit within the current scope of science, or there may be little reliable information available, or the values involved may lie outside of science. Moreover, although there may be at any one time a broad consensus on the bulk of scientific knowledge, the agreement does not extend to all scientific issues, let alone to all science-related social issues. And of course, on issues outside of their expertise, the opinions of scientists should enjoy no special credibility.

In their work, scientists go to great lengths to avoid bias—their own as well as that of others. But in matters of public interest, scientists, like other people, can be expected to be biased where their own personal, corporate, institutional, or community interests are at stake. For example, because of their commitment to science, many scientists may understandably be less than objective in their beliefs on how science is to be funded in comparison to other social needs.

C. A. Doxiadis, model of boulevard network, central Paris (twentieth century).

CHAPTER 2

THE NATURE OF MATHEMATICS

Mathematics relies on both logic and creativity, and it is pursued both for a variety of practical purposes and for its intrinsic interest. For some people, and not only professional mathematicians, the essence of mathematics lies in its beauty and its intellectual challenge. For others, including many scientists and engineers, the chief value of mathematics is how it applies to their own work. Because mathematics plays such a central role in modern culture, some basic understanding of the nature of mathematics is requisite for scientific literacy. To achieve this, students need to perceive mathematics as part of the scientific endeavor, comprehend the nature of mathematical thinking, and become familiar with key mathematical ideas and skills.

This chapter focuses on mathematics as part of the scientific endeavor and then on mathematics as a process, or way of thinking. Recommendations related to mathematical ideas are presented in Chapter 9, "The Mathematical World," and those on mathematical skills are included in Chapter 12, "Habits of Mind."

RECOMMENDATIONS

SOME FEATURES OF MATHEMATICS

Mathematics is the science of patterns and relationships. As a theoretical discipline, mathematics explores the possible relationships among abstractions without concern for whether those abstractions have counterparts in the real world. The abstractions can be anything from strings of numbers to geometric figures to sets of equations. In addressing, say, "Does the interval between prime numbers form a pattern?" as a theoretical question, mathematicians are interested only in finding a pattern or proving that there is none, but not in what use such knowledge might have. In deriving, for instance, an expression for the change in the surface area of any regular solid as its volume approaches zero, mathematicians have no interest in any correspondence between geometric solids and physical objects in the real world.

Mathematics is also an applied science. Many mathematicians focus their attention on solving problems that originate in the world of experience. They too search for patterns and relationships, and in the proc-ess they use techniques that are similar to those used in doing purely theoretical mathematics. The difference is largely one of intent. In contrast to theoretical mathematicians, applied mathematicians, in the examples given above, might study the interval pattern of prime numbers to develop a new system for coding numerical information, rather than as an abstract problem. Or they might tackle the area/volume problem as a step in producing a model for the study of crystal behavior.

The results of theoretical and applied mathematics often influence each other. The discoveries of theoretical mathematicians frequently turn out—sometimes decades later—to have unanticipated practical value. Studies on the mathematical properties of random events, for example, led to knowledge that later made it possible to improve the design of experiments in the social and natural sciences. Conversely, in trying to solve the problem of billing long-distance telephone users fairly, mathematicians made fundamental discoveries about the mathematics of complex networks. Theoretical mathematics, unlike the other sciences, is not constrained by the real world, but in the long run it contributes to a better understanding of that world.

Because of its abstractness, mathematics is universal in a sense that other fields of human thought are not. It finds useful applications in business, industry, music, historical scholarship, politics, sports, medicine, agriculture, engineering, and the social and natural sciences. The relationship between mathematics and the other fields of basic and applied science is especially strong. This is so for several reasons, including the following:

• The alliance between science and mathematics has a long history, dating back many centuries. Science provides mathematics with interesting problems to investigate, and mathematics provides science with powerful tools to use in analyzing data. Often, abstract patterns that have been studied for their own sake by mathematicians have turned out much later to be very useful in science. Science and mathematics are both trying to discover general patterns and relationships, and in this sense they are part of the same endeavor.

• Mathematics is the chief language of science. The symbolic language of mathematics has turned out to be extremely valuable for expressing scientific ideas unambiguously. The statement that $a = F/m$ is not simply a shorthand way of saying that the acceleration of an object depends on the force applied to it and its mass; rather, it is a precise statement of the quantitative relationship among those variables. More important, mathematics provides the grammar of science—the rules for analyzing scientific ideas and data rigorously.

• Mathematics and science have many features in common. These include a belief in understandable order; an interplay of imagination and rigorous logic; ideals of honesty and openness; the critical importance of peer criticism; the value placed on being the first to make a key discovery; being international in scope; and even, with the development of powerful electronic computers, being able to use technology to open up new fields of investigation.

• Mathematics and technology have also developed a fruitful relationship with each other. The mathematics of connections and logical chains, for example, has contributed greatly to the design of computer hardware and programing techniques. Mathematics also contributes more generally to engineering, as in describing complex systems whose behavior can then be simulated by computer. In those simulations, design features and operating conditions can be varied as a means of finding optimum designs. For its part, computer technology has opened up whole new areas in mathematics, even in the very nature of proof, and it also continues to help solve previously daunting problems.

MATHEMATICAL PROCESSES

Using mathematics to express ideas or to solve problems involves at least three phases: (1) representing some aspects of things abstractly, (2) manipulating the abstractions by rules of logic to find new relationships between them, and (3) seeing whether the new relationships say something useful about the original things.

Abstraction and Symbolic Representation

Mathematical thinking often begins with the process of abstraction—that is, noticing a similarity between two or more objects or events. Aspects that they have in common, whether concrete or hypothetical, can be represented by symbols such as numbers, letters, other marks, diagrams, geometrical constructions, or even words. Whole numbers are abstractions that represent the size of sets of things and events or the order of things within a set. The circle as a concept is an abstraction derived from human faces, flowers, wheels, or spreading ripples; the letter A may be an abstraction for the surface area of objects of any shape, for the acceleration of all moving objects, or for all objects having some specified property; the symbol + represents a process of addition, whether one is adding apples or oranges, hours, or miles per hour. And abstractions are made not only from concrete objects or processes; they can also be made from other abstractions, such as kinds of numbers (the even numbers, for instance).

Such abstraction enables mathematicians to concentrate on some features of things and relieves them of the need to keep other features continually in mind As far as mathematics is concerned, it does not matter whether a triangle represents the surface area of a sail or the convergence of two lines of sight on a star; mathematicians can work with either concept in the same way. The resulting economy of effort is very useful—provided that in making an abstraction, care is taken not to ignore features that play a significant role in determining the outcome of the events being studied.

Manipulating Mathematical Statements

After abstractions have been made and symbolic representations of them have been selected, those symbols become objects that can be combined and recombined in various ways according to precisely defined rules. Sometimes that is done with a fixed goal in mind; at other times it is done in the context of experiment or play to see what happens. Sometimes an appropriate manipulation can be identified easily from the intuitive meaning of the constituent words and symbols; at other times a useful series of manipulations has to be worked out by trial and error.

Typically, strings of symbols are combined into statements that express ideas or propositions. For example, the symbol A for the area of any square may be used with the symbol s for the length of the square's side to form the proposition $A = s^2$. This equation specifies how the area is related to the side—and also implies that it depends on nothing else. The rules of ordinary algebra can then be used to discover that if the length of the sides of a square is doubled, the square's area becomes four times as great. More generally, this knowledge makes it possible to find out what happens to the area of a square no matter how the length of its sides is changed, and conversely, how any change in the area affects the sides.

Mathematical insights into abstract relationships have grown over thousands of years, and they are still being extended—and sometimes revised. Although they began in the concrete experience of counting and measuring, they have come through many layers of abstraction and now depend much more on internal logic than on mechanical demonstration. In a sense, then, the manipulation of abstractions is much like a game: Start with some basic rules, then make any moves that fit those rules—which includes inventing additional rules and finding new connections between old rules. The test for the validity of new ideas is whether they are consistent and whether they relate logically to the other rules.

A central line of investigation in theoretical mathematics is identifying in each field of study a small set of basic ideas and rules from which all other interesting ideas and rules in that field can be logically deduced. Mathematicians, like other scientists, are particularly pleased when previously unrelated parts of mathematics are found to be derivable from one another, or from some more general theory. Part of the sense of beauty that many people have perceived in mathematics lies not in finding the greatest elaborateness or complexity but on the contrary, in finding the greatest economy and simplicity of representation and proof. As mathematics has progressed, more and more relationships have been found between parts of it that have been developed separately—for example, between the symbolic representations of algebra and the spatial representations of geometry. These cross-connections enable insights to be developed into the various parts; together, they strengthen belief in the correctness and underlying unity of the whole structure.

Application

Mathematical processes can lead to a kind of model of a thing, from which insights can be gained about the thing itself. Any mathematical relationships arrived at by manipulating abstract statements may or may not convey something truthful about the thing being modeled. For example, if 2 cups of water are added to 3 cups of water and the abstract mathematical operation $2 + 3 = 5$ is used to calculate the total, the correct answer is 5 cups of water. However, if 2 cups of sugar are added to 3 cups of hot tea and the same operation is used, 5 is an incorrect answer, for such an addition actually results in only slightly more than 4 cups of very sweet tea. The simple addition of volumes is appropriate to the first situation but not to the second—something that could have been predicted only by knowing something of the physical differences in the two situations. To be able to use and interpret mathematics well, therefore, it is necessary to be concerned with more than the mathematical validity of abstract operations and to also take into account how well they correspond to the properties of the things represented.

Sometimes common sense is enough to enable one to decide whether the results of the mathematics are appropriate. For example, to estimate the height 20 years from now of a girl who is 5′5″ tall and growing at the rate of an inch per year, common sense suggests rejecting the simple "rate times time" answer of 7′1″ as highly unlikely, and turning instead to some other mathematical model, such as curves that approach limiting values. Sometimes, however, it may be difficult to know just how appropriate mathematical results are—for example, when trying to predict stock-market prices or earthquakes.

Often a single round of mathematical reasoning does not produce satisfactory con-

clusions, and changes are tried in how the representation is made or in the operations themselves. Indeed, jumps are commonly made back and forth between steps, and there are no rules that determine how to proceed. The process typically proceeds in fits and starts, with many wrong turns and dead ends. This process continues until the results are good enough.

But what degree of accuracy is good enough? The answer depends on how the result will be used, on the consequences of error, and on the likely cost of modeling and computing a more accurate answer. For example, an error of 1 percent in calculating the amount of sugar in a cake recipe could be unimportant, whereas a similar degree of error in computing the trajectory for a space probe could be disastrous. The importance of the "good enough" question has led, however, to the development of mathematical processes for estimating how far off results might be and how much computation would be required to obtain the desired degree of accuracy.

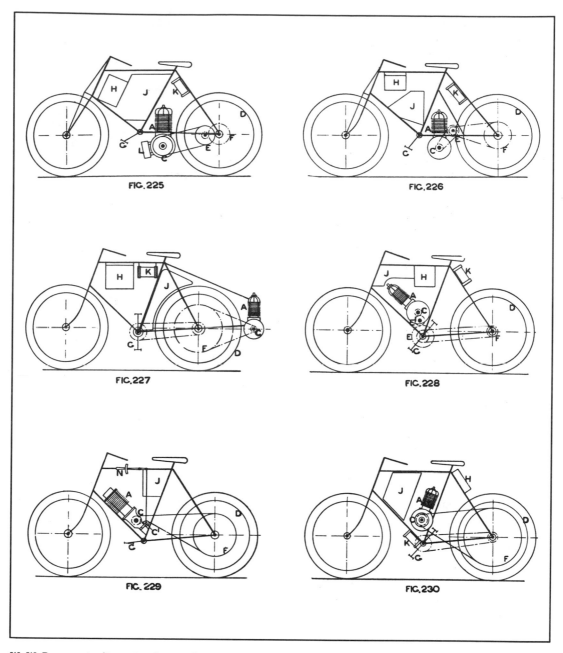

FIG. 225   FIG. 226

FIG. 227   FIG. 228

FIG. 229   FIG. 230

W. W. Beaumont, alternative designs for motorized bicycles (1903).

CHAPTER 3

THE NATURE OF TECHNOLOGY

As long as there have been people, there has been technology. Indeed, the techniques of shaping tools are taken as the chief evidence of the beginning of human culture. On the whole, technology has been a powerful force in the development of civilization, all the more so as its link with science has been forged. Technology—like language, ritual, values, commerce, and the arts—is an intrinsic part of a cultural system and it both shapes and reflects the system's values. In today's world, technology is a complex social enterprise that includes not only research, design, and crafts but also finance, manufacturing, management, labor, marketing, and maintenance.

In the broadest sense, technology extends our abilities to change the world: to cut, shape, or put together materials; to move things from one place to another; to reach farther with our hands, voices, and senses. We use technology to try to change the world to suit us better. The changes may relate to survival needs such as food, shelter, or defense, or they may relate to human aspirations such as knowledge, art, or control. But the results of changing the world are often complicated and unpredictable. They can include unexpected benefits, unexpected costs, and unexpected risks—any of which may fall on different social groups at different times. Anticipating the effects of technology is therefore as important as advancing its capabilities.

This chapter presents recommendations on what knowledge about the nature of technology is required for scientific literacy and emphasizes ways of thinking about technology that can contribute to using it wisely. The ideas are sorted into three sections: the connection of science and technology, the principles of technology itself, and the connection of technology and society. Chapter 8, "The Designed World," presents principles relevant to some of the key technologies of today's world. Chapter 10, "Historical Perspectives," includes a discussion of the Industrial Revolution. Chapter 12, "Habits of Mind," includes some skills relevant to participating in a technological world.

RECOMMENDATIONS

SCIENCE AND TECHNOLOGY

Technology Draws on Science and Contributes to It

In earlier times, technology grew out of personal experience with the properties of things and with the techniques for manipulating them, out of know-how handed down from experts to apprentices over many generations. The know-how handed down today is not only the craft of single practitioners but also a vast literature of words, numbers, and pictures that describe and give directions. But just as important as accumulated practical knowledge is the contribution to technology that comes from understanding the principles that underlie how things behave—that is, from scientific understanding.

Engineering, the systematic application of scientific knowledge in developing and applying technology, has grown from a craft to become a science in itself. Scientific knowledge provides a means of estimating what the behavior of things will be even before we make them or observe them. Moreover, science often suggests new kinds of behavior that had not even been imagined before, and so leads to new technologies. Engineers use knowledge of science and technology, together with strategies of design, to solve practical problems.

In return, technology provides the eyes and ears of science—and some of the muscle, too. The electronic computer, for example, has led to substantial progress in the study of weather systems, demographic patterns, gene structure, and other complex systems that would not have been possible otherwise. Technology is essential to science for purposes of measurement, data collection, treatment of samples, computation, transportation to research sites (such as Antarctica, the moon, and the ocean floor), sample collection, protection from hazardous materials, and communication. More and more, new instruments and techniques are being developed through technology that make it possible to advance various lines of scientific research.

Technology does not just provide tools for science, however; it also may provide motivation and direction for theory and research. The theory of the conservation of energy, for example, was developed in large part because of the technological problem of increasing the efficiency of commercial steam engines. The mapping of the locations of the entire set of genes in human DNA has been motivated by the technology of genetic engineering, which both makes such mapping possible and provides a reason for doing so.

As technologies become more sophisticated, their links to science become stronger. In some fields, such as solid-state physics (which involves transistors and superconductors), the ability to make something and the ability to study it are so interdependent that science and engineering can scarcely be separated. New technology often requires new understanding; new investigations often require new technology.

Engineering Combines Scientific Inquiry and Practical Values

The component of technology most closely allied to scientific inquiry and to mathematical modeling is engineering. In its broadest sense, engineering consists of construing a problem and designing a solution for it. The basic method is to first devise a general approach and then work out the technical details of the construction of requisite objects (such as an automobile engine, a computer chip, or a mechanical toy) or processes (such as irrigation, opinion polling, or product testing).

Much of what has been said about the nature of science applies to engineering as well, particularly the use of mathematics, the interplay of creativity and logic, the eagerness to be original, the variety of people involved, the professional specialties, public responsibility, and so on. Indeed, there are more people called engineers than people called scientists, and many scientists are doing work that could be described as engineering as well as science. Similarly, many engineers are engaged in science.

Scientists see patterns in phenomena as making the world understandable; engineers also see them as making the world manipulable. Scientists seek to show that theories fit the data; mathematicians seek to show logical proof of abstract connections; engineers seek to demonstrate that designs work. Scientists cannot provide answers to all questions; mathematicians cannot prove all possible connections; engineers cannot design solutions for all problems.

But engineering affects the social system and culture more directly than scientific research, with immediate implications for the success or failure of human enterprises and for personal benefit and harm. Engineering decisions, whether in designing an airplane bolt or an irrigation system, inevitably involve social and personal values as well as scientific judgments.

PRINCIPLES OF TECHNOLOGY

The Essence of Engineering Is Design Under Constraint

Every engineering design operates within constraints that must be identified and taken into account. One type of constraint is absolute—for example, physical laws such as the conservation of energy or physical properties such as limits of flexibility, electrical conductivity, and friction. Other types have some flexibility: economic (only so much money is available for this purpose), political (local, state, and national regulations), social (public opposition), ecological (likely disruption of the natural environment), and ethical (disadvantages to some people, risk to subsequent generations). An optimum design takes into account all the constraints and strikes some reasonable compromise among them. Reaching such design compromises—including, sometimes, the decision not to develop a particular technology further—requires taking personal and social values into account. Although design may sometimes require only routine decisions about the combining of familiar components, often it involves great creativity in inventing new approaches to problems, new components, and new combinations—and great innovation in seeing new problems or new possibilities.

But there is no perfect design. Accommodating one constraint well can often lead to conflict with others. For example, the lightest material may not be the strongest, or the most efficient shape may not be the safest or the most aesthetically pleasing. Therefore every design problem lends itself to many alternative solutions, depending on what values people place on the various constraints. For example, is strength more desirable than lightness, and is appearance more important than safety? The task is to arrive at a design that reasonably balances the many trade-offs, with the understanding that no single design is ever simultaneously the safest, the most reliable, the most efficient, the most inexpensive, and so on.

It is seldom practical to design an isolated object or process without considering the broad context in which it will be used. Most products of technology have to be operated, maintained, occasionally repaired, and ultimately replaced. Because all these related activities bear costs, they too have to be considered. A similar issue that is becoming increasingly important with more complex technologies is the need to train personnel to sell, operate, maintain, and repair them. Particularly when technology changes quickly, training can be a major cost. Thus, keeping down demands on personnel may be another design constraint.

Designs almost always require testing, especially when the design is unusual or complicated, when the final product or process is likely to be expensive or dangerous, or when failure has a very high cost. Performance tests of a design may be conducted by using complete products, but doing so may be prohibitively difficult or expensive. So testing is often done by using small-scale physical models, computer simulations, analysis of analogous systems (for example, laboratory animals standing in for humans, earthquake disasters for nuclear disasters), or testing of separate components only.

All Technologies Involve Control

All systems, from the simplest to the most complex, require control to keep them operating properly. The essence of control is comparing information about what is happening with what we want to happen and then making appropriate adjustments. Control typically requires feedback (from sensors or other sources of information) and logical comparisons of that information to instructions (and perhaps to other data input)—and a means for activating changes. For example, a baking oven is a fairly simple system that compares the information from a temperature sensor to a control setting and turns the heating element up or down to keep the temperature within a small range. An automobile is a more complex system, made up of subsystems for controlling engine temperature, combustion rate, direction, speed, and so forth, and for changing them when the immediate circumstances or instructions change. Miniaturized electronics makes possible logical control in a great variety of technical systems. Almost all but the simplest household appliances used today include microprocessors to control their performance.

As controls increase in complexity, they too require coordination, which means additional layers of control. Improvement in rapid communication and rapid processing of information makes possible very elaborate systems of control. Yet all technological systems include human as well as mechanical or electronic components. Even the most automatic system requires human control at some point—to program the built-in control elements, monitor them, take over from them when they malfunction, and change them when the purposes of the system change. The ultimate control lies with people who understand in some depth what the purpose and nature of the control process are and the context within which the process operates.

Technologies Always Have Side Effects

In addition to its intended benefits, every design is likely to have unintended side effects in its production and application. On the one hand, there may be unexpected benefits. For example, working conditions may become safer when materials are molded rather than stamped, and materials designed for space satellites may prove useful in consumer products. On the other hand, substances or processes involved in production may harm production workers or the public in general; for example, sitting in front of a computer may strain the user's eyes and lead to isolation from other workers. And jobs may be affected—by increasing employment for people involved in the new technology, decreasing employment for others involved in the old technology, and changing the nature of the work people must do in their jobs.

It is not only large technologies—nuclear reactors or agriculture—that are prone to side effects, but also the small, everyday ones. The effects of ordinary technologies may be individually small but collectively significant. Refrigerators, for example, have had a predictably favorable impact on diet and on food distribution systems. Because there are so many refrigerators, however, the tiny leakage of a gas used in their cooling systems may have substantial adverse effects on the earth's atmosphere.

Some side effects are unexpected because of a lack of interest or resources to predict them. But many are not predictable even in principle because of the sheer complexity of technological systems and the inventiveness of people in finding new ap-

plications. Some unexpected side effects may turn out to be ethically, aesthetically, or economically unacceptable to a substantial fraction of the population, resulting in conflict between groups in the community. To minimize such side effects, planners are turning to systematic risk analysis. For example, many communities require by law that environmental impact studies be made before they will consider giving approval for the introduction of a new hospital, factory, highway, waste-disposal system, shopping mall, or other structure.

Risk analysis, however, can be complicated. Because the risk associated with a particular course of action can never be reduced to zero, acceptability may have to be determined by comparison to the risks of alternative courses of action, or to other, more familiar risks. People's psychological reactions to risk do not necessarily match straightforward mathematical models of benefits and costs. People tend to perceive a risk as higher if they have no control over it (smog versus smoking) or if the bad events tend to come in dreadful peaks (many deaths at once in an airplane crash versus only a few at a time in car crashes). Personal interpretation of risks can be strongly influenced by how the risk is stated—for example, comparing the probability of dying versus the probability of surviving, the dreaded risks versus the readily acceptable risks, the total costs versus the costs per person per day, or the actual number of people affected versus the proportion of affected people.

All Technological Systems Can Fail

Most modern technological systems, from transistor radios to airliners, have been engineered and produced to be remarkably reliable. Failure is rare enough to be surprising. Yet the larger and more complex a system is, the more ways there are in which it can go wrong—and the more widespread the possible effects of failure. A system or device may fail for different reasons: because some part fails, because some part is not well matched to some other, or because the design of the system is not adequate for all the conditions under which it is used. One hedge against failure is overdesign—that is, for example, making something stronger or bigger than is likely to be necessary. Another hedge is redundancy—that is, building in one backup system or more to take over in case the primary one fails.

If failure of a system would have very costly consequences, the system may be designed so that its most likely way of failing would do the least harm. Examples of such "fail-safe" designs are bombs that cannot explode when the fuse malfunctions; automobile windows that shatter into blunt, connected chunks rather than into sharp, flying fragments; and a legal system in which uncertainty leads to acquittal rather than conviction. Other means of reducing the likelihood of failure include improving the design by collecting more data, accommodating more variables, building more realistic working models, running computer simulations of the design longer, imposing tighter quality control, and building in controls to sense and correct problems as they develop.

All of the means of preventing or minimizing failure are likely to increase cost. But no matter what precautions are taken or resources invested, risk of technological failure can never be reduced to zero. Analysis of risk, therefore, involves estimating a probability of occurrence for every undesirable outcome that can be foreseen—and also estimating a measure of the harm that would be done if it did occur. The expected importance of each risk is then estimated by combining its probability and its measure of harm. The relative risk of different designs can then be compared in terms of the combined probable harm resulting from each.

TECHNOLOGY AND SOCIETY

Technological and Social Systems Interact Strongly

Individual inventiveness is essential to technological innovation. Nonetheless, social and economic forces strongly influence what technologies will be undertaken, paid attention to, invested in, and used. Such decisions occur directly as a matter of government policy and indirectly as a consequence of the circumstances and values of a society at any particular time. In the United States, decisions about which technological options will prevail are influenced by many factors, such as consumer acceptance, patent laws, the availability of risk capital, the federal budget process, local and national regulations, media attention, economic competition, tax incentives, and scientific discoveries. The balance of such incentives and regulations usually bears differently on different technological systems, encouraging some and discouraging others.

Technology has strongly influenced the course of history and the nature of human society, and it continues to do so. The great revolutions in agricultural technology, for example, have probably had more influence on how people live than political revolutions; changes in sanitation and preventive medicine have contributed to the population explosion (and to its control); bows and arrows, gunpowder, and nuclear explosives have in their turn changed how war is waged; and the microprocessor is changing how people write, compute, bank, operate businesses, conduct research, and communicate with one another. Technology is largely responsible for such large-scale changes as the increased urbanization of society and the dramatically growing economic interdependence of communities worldwide.

Historically, some social theorists have believed that technological change (such as industrialization and mass production) causes social change, whereas others have believed that social change (such as political or religious changes) leads to technological change. However, it is clear that because of the web of connections between technological and other social systems, many influences act in both directions.

The Social System Imposes Some Restrictions on Openness in Technology

For the most part, the professional values of engineering are very similar to those of science, including the advantages seen in the open sharing of knowledge. Because of the economic value of technology, however, there are often constraints on the openness of science and engineering that are relevant to technological innovation. A large investment of time and money and considerable commercial risk are often required to develop a new technology and bring it to market. That investment might well be jeopardized if competitors had access to the new technology without making a similar investment, and hence companies are often reluctant to share technological knowledge. But no scientific or technological knowledge is likely to remain secret for very long. Secrecy most often provides only an advantage in terms of time—a head start, not absolute control of knowledge. Patent laws encourage openness by giving individuals and companies control over the use of any new technology they develop; however, to promote technological competition, such control is only for a limited period of time.

Commercial advantage is not the only motivation for secrecy and control. Much technological development occurs in settings, such as government agencies, in which commercial concerns are minimal but national security concerns may lead to secrecy. Any technology that has potential military applications can arguably be subject to restrictions imposed by the federal government, which may limit the sharing of engineering knowledge—or even the exportation of products from which engineering knowledge could be inferred. Because the connections between science and technology are so close in some fields, secrecy inevitably begins to restrict some of the free flow of information in science as well. Some scientists and engineers are very uncomfortable with what they perceive as a compromise of the scientific ideal, and some refuse to work on projects that impose secrecy. Others, however, view the restrictions as appropriate.

Decisions About the Use of Technology Are Complex

Most technological innovations spread or disappear on the basis of free-market forces—that is, on the basis of how people and companies respond to such innovations. Occasionally, however, the use of some technology becomes an issue subject to public debate and possibly formal regulation. One way in which technology becomes such an issue is when a person, group, or business proposes to test or introduce a new technology—as has been the case with contour plowing, vaccination, genetic engineering, and nuclear power plants. Another way is when a technology already in widespread use is called into question—as, for example, when people are told (by

individuals, organizations, or agencies) that it is essential to stop or reduce the use of a particular technology or technological product that has been discovered to have, or that may possibly have, adverse effects. In such instances, the proposed solution may be to ban the burial of toxic wastes in community dumps, or to prohibit the use of leaded gasoline and asbestos insulation.

Rarely are technology-related issues simple and one-sided. Relevant technical facts alone, even when known and available (which often they are not), usually do not settle matters entirely in favor of one side or the other. The chances of reaching good personal or collective decisions about technology depend on having information that neither enthusiasts nor skeptics are always ready to volunteer. The long-term interests of society are best served, therefore, by having processes for ensuring that key questions concerning proposals to curtail or introduce technology are raised and that as much relevant knowledge as possible is brought to bear on them. Considering these questions does not ensure that the best decision will always be made, but the failure to raise key questions will almost certainly result in poor decisions. The key questions concerning any proposed new technology should include the following:

• What are alternative ways to accomplish the same ends? What advantages and disadvantages are there to the alternatives? What trade-offs would be necessary between positive and negative side effects of each?

• Who are the main beneficiaries? Who will receive few or no benefits? Who will suffer as a result of the proposed new technology? How long will the benefits last? Will the technology have other applications? Whom will they benefit?

• What will the proposed new technology cost to build and operate? How does that compare to the cost of alternatives? Will people other than the beneficiaries have to bear the costs? Who should underwrite the development costs of a proposed new technology? How will the costs change over time? What will the social costs be?

• What risks are associated with the proposed new technology? What risks are associated with not using it? Who will be in greatest danger? What risk will the technology present to other species of life and to the environment? In the worst possible case, what trouble could it cause? Who would be held responsible? How could the trouble be undone or limited?

• What people, materials, tools, knowledge, and know-how will be needed to build, install, and operate the proposed new technology? Are they available? If not, how will they be obtained, and from where? What energy sources will be needed for construction or manufacture, and also for operation? What resources will be needed to maintain, update, and repair the new technology?

• What will be done to dispose safely of the new technology's waste materials? As it becomes obsolete or worn out, how will it be replaced? And finally, what will become of the material of which it was made and the people whose jobs depended on it?

Individual citizens may seldom be in a position to ask or demand answers for these questions on a public level, but their knowledge of the relevance and importance of answers increases the attention given to the questions by private enterprise, interest groups, and public officials. Furthermore, individuals may ask the same questions with regard to their own use of technology—for instance, their own use of efficient household appliances, of substances that contribute to pollution, of foods and fabrics. The cumulative effect of individual decisions can have as great an impact on the large-scale use of technology as pressure on public decisions can.

Not all such questions can be answered readily. Most technological decisions have to be made on the basis of incomplete information, and political factors are likely to have as much influence as technical ones, and sometimes more. But scientists, mathematicians, and engineers have a special role in looking as far ahead and as far afield as is practical to estimate benefits, side effects, and risks. They can also assist by designing adequate detection devices and monitoring techniques, and by setting up procedures for the collection and statistical analysis of relevant data.

Thomas Hart Benton, *Trail Riders* (1964–65).

CHAPTER 4

THE PHYSICAL SETTING

Humans have never lost interest in trying to find out how the universe is put together, how it works, and where they fit in the cosmic scheme of things. The development of our understanding of the architecture of the universe is surely not complete, but we have made great progress. Given a universe that is made up of distances too vast to reach and of particles too small to see and too numerous to count, it is a tribute to human intelligence that we have made as much progress as we have in accounting for how things fit together. All humans should participate in the pleasure of coming to know their universe better.

This chapter consists of recommendations for basic knowledge about the overall structure of the universe and the physical principles on which it seems to run, with emphasis on the earth and the solar system. The chapter focuses on two principal subjects: the structure of the universe and the major processes that have shaped the planet earth, and the concepts with which science describes the physical world in general—organized for convenience under the headings of matter, energy, motion, and forces.

RECOMMENDATIONS

THE UNIVERSE

The universe is large and ancient, on scales staggering to the human mind. The earth has existed for only about a third of the history of the universe and is in comparison a mere speck in space. Our sun is a medium-sized star orbiting near the edge of the arm of an ordinary disk-shaped galaxy of stars, part of which we can see as a vast glowing band that spans the sky on a clear night (the Milky Way). Our galaxy contains many billion stars, and the universe contains many billion such galaxies, some of which we may be able to see with the naked eye as fuzzy spots on a clear night.

Using our fastest rockets, it would still take us thousands of years to reach the star nearest our sun. Even light from that nearest star takes four years to reach us. And the light reaching us from the farthest galaxies left them at a time not long after the beginning of the universe. That is why when we observe the stars, we are observing their past.

There are wondrously different kinds of stars that are much larger and much smaller, much hotter and much cooler, much older and much younger than our sun. Most of them apparently are not an isolated single star as our sun is but are part of systems of two or more stars orbiting around a common center of mass. So too there are other galaxies and clusters of galaxies different from our own in size, shape, and direction of motion. But in spite of this variety, they all appear to be composed of the same elements, forces, and forms of energy found in our own solar system and galaxy, and they appear to behave according to the same physical principles.

It seems that the entire contents of the known universe expanded explosively into existence from a single hot, dense, chaotic mass more than ten billion years ago. Stars coalesced out of clouds of the lightest elements (hydrogen and helium), heated up from the energy of falling together, and began releasing nuclear energy from the fusion of light elements into heavier ones in their extremely hot, dense cores. Eventually, many of the stars exploded, producing new clouds from which other stars—and presumably planets orbiting them—could condense. The process of star formation continues. Stars are formed and eventually dissipate, and matter and energy change forms—as they have for billions of years.

Our solar system coalesced out of a giant cloud of gas and debris left in the wake of exploding stars about five billion years ago. Everything in and on the earth, including living organisms, is made of this material. As the earth and the other planets formed, the heavier elements fell to their centers. On planets close to the sun (Mercury, Venus, Earth, and Mars), the lightest elements were mostly blown or boiled away by radiation from the newly formed sun; on the outer planets (Jupiter, Saturn, Uranus, Neptune, and Pluto), the lighter elements still surround them as deep atmospheres of gas or as frozen solid layers.

In total, there are nine planets of very different size, composition, and surface features that move around the sun in nearly circular orbits. Around the planets orbit a great variety of moons and (in some cases) flat rings of rock and ice debris or (in the case of the earth) a moon and artificial satellites. Features of many of the planets and their moons show evidence of developmental processes similar to those that occur on the earth (such as earthquakes, lava flows, and erosion).

There are also a great many smaller bodies of rock and ice orbiting the sun. Some of those that the earth encounters in its yearly orbit around the sun burn up from friction as they plunge into the atmosphere. Some have such long and off-center orbits that they periodically come very close to the sun, whose radiation boils off material and pushes it into a long illuminated tail that we see as a comet.

Our still-growing knowledge of the solar system and the rest of the universe comes to us in part by direct observation but mostly through the use of tools we have developed to extend and supplement our own senses. These tools include radio and x-ray telescopes that are sensitive to a broad spectrum of information coming to us from space; computers that can undertake increasingly complicated calculations of gravitational systems or nuclear reactions, finding patterns in data and deducing the implications

of theories; space probes that send back detailed pictures and other data from distant planets in our own solar system; and huge "atom smashers" that simulate conditions in the early universe and probe the inner workings of atoms.

Most of what we believe we know about the universe must be inferred by using all these tools to look at very small slices of space and time. What we know about stars is based on analysis of the light that reaches us from them. What we know about the interior of the earth is based on measurements we make on or near its surface or from satellites orbiting above the surface. What we know about the evolution of the sun and planets comes from studying the radiation from a small sample of stars, visual features of the planets, and samples of material (such as rock, meteorites, and moon and Mars scrapings), and imagining how they got to be the way they are.

THE EARTH

We live on a fairly small planet, the third from the sun in the only system of planets definitely known to exist (although similar systems are likely to be common in the universe). Like that of all planets and stars, the earth's shape is approximately spherical, the result of mutual gravitational attraction pulling its material toward a common center. Unlike the much larger outer planets, which are mostly gas, the earth is mostly rock, with three-fourths of its surface covered by a relatively thin layer of water and the entire planet enveloped by a thin blanket of air. Bulges in the water layer are raised on both sides of the planet by the gravitational tugs of the moon and sun, producing high tides about twice a day along ocean shores. Similar bulges are produced in the blanket of air as well.

Of all the diverse planets and moons in our solar system, only the earth appears to be capable of supporting life as we know it. The gravitational pull of the planet's mass is sufficient to hold onto an atmosphere. This thin envelope of gases evolved as a result of changing physical conditions on the earth's surface and the evolution of plant life, and it is an integral part of the global ecosystem. Altering the concentration of the natural component gases of the atmosphere, or adding new ones, can have serious consequences for the earth's life systems.

The distance of the earth from the sun ensures that energy reaches the planet at a rate sufficient to sustain life, and yet not so fast that water would boil away or that molecules necessary to life would not form. Water exists on the earth in liquid, solid, and gaseous forms—a rarity among the planets (the others are either closer to the sun and too hot or farther from the sun and too cold).

The motion of the earth and its position with regard to the sun and the moon have noticeable effects. The earth's one-year revolution around the sun, because of the tilt of the earth's axis, changes how directly sunlight falls on one part or another of the earth. This difference in heating different parts of the earth's surface produces seasonal variations in climate. The rotation of the planet on its axis every 24 hours produces the planet's night-and-day cycle—and (to observers on earth) makes it seem as though the sun, planets, stars, and moon are orbiting the earth. The combination of the earth's motion and the moon's own orbit around the earth, once in about 28 days, results in the phases of the moon (on the basis of the changing angle at which we see the sunlit side of the moon).

The earth has a variety of climatic patterns, which consist of different conditions of temperature, precipitation, humidity, wind, air pressure, and other atmospheric phenomena. These patterns result from an interplay of many factors. The basic energy source is the heating of land, ocean, and air by solar radiation. Transfer of heat energy at the interfaces of the atmosphere with the land and oceans produces layers at different temperatures in both the air and the oceans. These layers rise or sink or mix, giving rise to winds and ocean currents that carry heat energy between warm and cool regions. The earth's rotation curves the flow of winds and ocean currents, which are further deflected by the shape of the land.

The cycling of water in and out of the atmosphere plays an important part in determining climatic patterns—evaporating from the surface, rising and cooling, condensing into clouds and then into snow or rain, and falling again to the surface, where it collects in rivers, lakes, and porous layers

of rock. There are also large areas on the earth's surface covered by thick ice (such as Antarctica), which interacts with the atmosphere and oceans in affecting worldwide variations in climate.

The earth's climates have changed radically and they are expected to continue changing, owing mostly to the effects of geological shifts such as the advance or retreat of glaciers over centuries of time or a series of huge volcanic eruptions in a short time. But even some relatively minor changes of atmospheric content or of ocean temperature, if sustained long enough, can have widespread effects on climate.

The earth has many resources of great importance to human life. Some are readily renewable, some are renewable only at great cost, and some are not renewable at all. The earth comprises a great variety of minerals, whose properties depend on the history of how they were formed as well as on the elements of which they are composed. Their abundance ranges from rare to almost unlimited. But the difficulty of extracting them from the environment is as important an issue as their abundance. A wide variety of minerals are sources for essential industrial materials, such as iron, aluminum, magnesium, and copper. Many of the best sources are being depleted, making it more and more difficult and expensive to obtain those minerals.

Fresh water is an essential resource for daily life and industrial processes. We obtain our water from rivers and lakes and from water that moves below the earth's surface. This groundwater, which is a major source for many people, takes a long time to accumulate in the quantities now being used. In some places it is being depleted at a very rapid rate. Moreover, many sources of fresh water cannot be used because they have been polluted.

Wind, tides, and solar radiation are continually available and can be harnessed to provide sources of energy. The oceans, atmosphere, topsoil, sea creatures, and trees are also renewable resources. However, it can be enormously expensive to clean up polluted air and water, restore destroyed forests and fishing grounds, or restore or preserve eroded soils of poorly managed agricultural areas. Although the oceans and atmosphere are very large and have a great capacity to absorb and recycle materials naturally, they do have their limits. They have only a finite capacity to withstand change without generating major ecological alterations that may also have adverse effects on human activities.

FORCES THAT SHAPE THE EARTH

The interior of the earth is hot, under high pressure from the weight of overlying layers, and more dense than its rocky crust. Forces within the earth cause continual changes on its surface. The solid crust of the earth—including both the continents and ocean basins—consists of separate sections that overlie a hot, almost molten layer. The separate crustal plates move on this softer layer—as much as an inch or more per year—colliding in some places, pulling apart in others. Where the crustal plates collide, they may scrape sideways, or compress the land into folds that eventually become mountain ranges (such as the Rocky Mountains and the Himalayas); or one plate may slide under the other and sink deeper into the earth. Along the boundaries between colliding plates, earthquakes shake and break the surface, and volcanic eruptions release molten rock from below, also building up mountains.

Where plates separate under continents, the land sinks to form ever-widening valleys. When separation occurs in the thin regions of plates that underlie ocean basins, molten rock wells up to create ever-wider ocean floors. Volcanic activity along these mid-ocean separations may build up undersea mountains that are far higher than those rising from the land surface—sometimes thrusting above the water's surface to create mid-ocean islands (such as Hawaii). Minerals are made, dissolved, and remade—on the earth's surface, in the oceans, and in the hot, high-pressure layers beneath the crust.

Waves, wind, water, and ice sculpt the earth's surface to produce distinctive landforms. Rivers and glacial ice carry off soil and break down rock, eventually depositing the material in sediments or carrying it in solution to the sea. Some of these effects occur rapidly and others very slowly. For instance, many of the features of the earth's surface today can be traced to the motion of glaciers back and forth across much of the

northern hemisphere over a period lasting more than a million years. By contrast, the shoreline can change almost overnight—as waves erode the shores, and wind carries off loose surface material and deposits it elsewhere.

Elements such as carbon, oxygen, nitrogen, and sulfur cycle slowly through the land, oceans, and atmosphere, changing their locations and chemical combinations. Sediments of sand and shells of dead organisms are gradually buried, subjected to great pressure, and eventually turned into solid rock again. These re-formed rock layers may be forced up again to become land surface and eventually even mountains. Thousands upon thousands of layers of sedimentary rock testify to the long history of the earth, and to the long history of changing life forms whose remains are found in successive layers of rock.

Plants and animals reshape the landscape in many ways. The composition and texture of the soil, and consequently its fertility and resistance to erosion, are greatly influenced by plant roots and debris, bacteria, fungi, worms, rodents, and other animals, as they break up the soil and add organic material to it. The presence of life has also altered the earth's atmosphere. Plants remove carbon dioxide from the air, use the carbon for synthesizing sugars, and release oxygen. This process is responsible for the oxygen in our air today.

The landforms, climate, and resources of the earth's surface affect where and how people live and how human history has unfolded. At the same time, human activities have changed the earth's land surface, oceans, and atmosphere. For instance, reducing the amount of forest cover on the earth's surface has led to a dramatic increase in atmospheric carbon dioxide, which in turn has led to an increase in the average temperature of the earth's atmosphere and surface. Smoke and other substances from human activity interact chemically with the atmosphere and produce undesirable effects such as smog, acid rain, and an increase in the damaging ultraviolet radiation that penetrates the atmosphere. Intensive farming has stripped land of vegetation and topsoil, creating virtual deserts in some parts of the world.

THE STRUCTURE OF MATTER

The things of the physical world seem to be made up of a stunningly varied array of materials. Materials differ greatly in shape, density, flexibility, texture, toughness, and color; in their ability to give off, absorb, bend, or reflect light; in what form they take at different temperatures; in their responses to each other; and in hundreds of other ways. Yet, in spite of appearances, everything is really made up of a relatively few kinds of basic material combined in various ways. As it turns out, about 100 such materials—the chemical elements—are now known to exist, and only a few of them are abundant in the universe. When two or more substances interact to form new substances (as in burning, digestion, corrosion, and cooking), the elements composing them combine in new ways. In such recombinations, the properties of the new combinations may be very different from those of the old.

The basic premise of the modern theory of matter is that the elements consist of a few different kinds of atoms—particles far too tiny to see in a microscope—that join together in different configurations to form substances. There are one or more—but never many—kinds of these atoms for each of the approximately 100 elements. Each atom is composed of a central, positively charged nucleus—only a very small fraction of the atom's volume, but containing most of its mass—surrounded by a cloud of much lighter, negatively charged electrons. The number of electrons in an atom—ranging from 1 up to about 100—matches the number of charged particles, or protons, in the nucleus, and determines how the atom will link to other atoms to form molecules. Electrically neutral particles (neutrons) in the nucleus add to its mass but do not affect the number of electrons and so have almost no effect on the atom's links to other atoms (its chemical behavior). A block of pure carbon, for instance, is made up of two kinds, or isotopes, of carbon atoms that differ somewhat in mass but have almost identical chemical properties.

Every substance can exist in a variety of different states, depending on temperature and pressure. Just as water can exist as ice, water, and vapor, all but a few substances can also take solid, liquid, and gaseous form. When matter gets cold enough, atoms or molecules lock in place in a more or less orderly fashion as solids. Increasing the temperature means increasing the average energy of motion of the atoms. So if the temperature is increased, atoms and molecules become more agitated and usually move slightly farther apart; that is, the material expands. At higher temperatures, the atoms and molecules are more agitated still and can slide past one another while remaining loosely bound, as in a liquid. At still higher temperatures, the agitation of the atoms and molecules overcomes the attractions between them and they can move around freely, interacting only when they happen to come very close—usually bouncing off one another, as in a gas. As temperature increases in either a liquid or gas, molecules collide more often and intermix more rapidly, and thus the rate of most chemical and physical reactions also increases.

As the temperature rises even higher, eventually the energy of collisions breaks all molecules apart into atoms, and knocks electrons away from atoms. At extremely high temperatures, the nuclei of atoms may get so close during collisions that they are affected by the strong internal nuclear forces, and nuclear reactions may occur.

The arrangement of the outermost electrons in an atom determines how the atom can bond to others and form materials. Bonds are formed between atoms when electrons are transferred from one atom to another, or when electrons are more or less shared between them. Depending on what kinds of bonds are made, the atoms may link together in chaotic mixtures, in distinctive molecules that have a uniform number and configuration of atoms, or in the symmetrically repeated patterns of crystal arrays. Molecular configurations can be as simple as pairs of identical atoms (such as oxygen molecules) or as complex as folded and cross-linked chains thousands of atoms long (such as protein and DNA molecules). The exact shapes of these complex molecules is a critical factor in how they interact with one another. Crystal arrays may be entirely regular, or permeated with irregularities of composition and structure. The small differences in composition and structure can give materials very different properties.

The configuration of electrons in atoms determines what reactions can occur between atoms, how much energy is required to get the reaction to happen, and how much energy is released in the reaction. The rates at which reactions occur in large collections of atoms depend largely on the immediate environment (such as temperature, pressure, and concentrations of the reacting substances). Reaction rates can be affected significantly by very small concentrations of some atoms and molecules. For example, some molecules participate in complicated reactions in ways that greatly increase or decrease the reaction rates but end up unchanged themselves. These catalyst molecules may link to the reactants in a way that positions them well to link to each other, or the molecules may have an excited state that can transfer just the right amount of energy for the reaction to occur.

Each of the elements that make up familiar substances consists of only a few naturally occurring isotopes. Most other possible isotopes of any element are unstable and, if they happen to be formed, sooner or later will decay into some isotope of another element (which may itself be unstable). The decay involves emission of particles and radiation from the nucleus—that is, radioactivity. In the materials of the earth, there are small proportions of some radioactive isotopes that were left over from the original formation of heavy elements in stars, that were formed more recently by impacts of nuclear particles from space, or that result from the nuclear decay of other isotopes. Together, these isotopes produce a low level of background radiation in the general environment.

It is not possible to predict when an unstable nucleus will decay. We can determine only what fraction of a collection of identical nuclei are likely to decay in a given period of time. The half-life of an unstable isotope is the time it takes for half of the nuclei in any sample of that isotope to decay; half-lives of different isotopes range from less than a millionth of a second to many millions of years. The half-life of any particular isotope is constant and unaffected by physical conditions such as pressure and temperature. Radioactivity can therefore be used to estimate the passage of time, by measuring the fraction of nuclei that have already decayed. For example, the fraction of an unstable, long-half-life isotope remaining in a sample of rock can be used to estimate how long ago the rock was formed.

TRANSFORMATIONS OF ENERGY

Energy appears in many forms, including radiation, the motion of bodies, excited states of atoms, and strain within and between molecules. All of these forms are in an important sense equivalent, in that one form can change into another. Most of what goes on in the universe—such as the collapsing and exploding of stars, biological growth and decay, the operation of machines and computers—involves one form of energy being transformed into another.

Forms of energy can be described in different ways: Sound energy is chiefly the regular back-and-forth motion of molecules; heat energy is the random motion of molecules; gravitational energy lies in the separation of mutually attracting masses; the energy stored in mechanical strains involves the separation of mutually attracting electric charges. Although the various forms appear very different, each can be measured in a way that makes it possible to keep track of how much of one form is converted into another. Whenever the amount of energy in one place or form diminishes, the amount in another place or form increases by an equivalent amount. Thus, if no energy leaks in or out across the boundaries of a system, the total energy of all the different forms in the system will not change, no matter what kinds of gradual or violent changes actually occur within the system.

But energy does tend to leak across boundaries. In particular, transformations of energy usually result in producing some energy in the form of heat, which leaks away by radiation or conduction (such as from engines, electrical wires, hot-water tanks, our bodies, and stereo systems). Thus, the total amount of energy available for transformation is almost always decreasing. For example, almost all of the energy stored in the molecules of gasoline used during an automobile trip goes, by way of friction and exhaust, into producing a slightly warmer car, road, and air. But even if such diffused energy is prevented from leaking away, it tends to distribute itself evenly and thus may no longer be useful to us. This is because energy can accomplish transformations only when it is concentrated more in some places than in others (such as in falling water, in high-energy molecules in fuels and food, in unstable nuclei, and in radiation from the intensely hot sun). When energy is transformed into heat energy that diffuses all over, further transformations are less likely.

The reason that heat tends always to diffuse from warmer places to cooler places is a matter of probability. Heat energy in a material consists of the disordered motions of its perpetually colliding atoms or molecules. As very large numbers of atoms or molecules in one region of a material repeatedly and randomly collide with those of a neighboring region, there are far more ways in which their energy of random motion can end up shared about equally throughout both regions than there are ways in which it can end up more concentrated in one region. The disordered sharing of heat energy all over is therefore far more likely to occur than any more orderly concentration of heat energy in any one place. More generally, in any interactions of atoms or molecules, the statistical odds are that they will end up in more disorder than they began with.

It is, however, entirely possible for some systems to increase in orderliness—as long as systems connected to them increase even more in disorderliness. The cells of a human organism, for example, are always busy increasing order, as in building complex molecules and body structures. But this occurs at the cost of increasing the disorder around us even more—as in breaking down the molecular structure of food we eat and in warming up our surroundings. The point is that the *total* amount of disorder always tends to increase.

Different energy levels are associated with different configurations of atoms in molecules. Some changes in configuration require additional energy, whereas other changes release energy. For example, heat energy has to be supplied to start a charcoal fire (by evaporating some carbon atoms away from others in the charcoal); however, when oxygen molecules combine with the carbon atoms into the lower-energy configuration of a carbon dioxide molecule, much more energy is released as heat and

light. Or a chlorophyll molecule can be excited to a higher-energy configuration by sunlight; the chlorophyll in turn excites molecules of carbon dioxide and water so they can link, through several steps, into the higher-energy configuration of a molecule of sugar (plus some regenerated oxygen). Later, the sugar molecule may subsequently interact with oxygen to yield carbon dioxide and water molecules again, transferring the extra energy from sunlight to still other molecules.

It becomes evident, at the molecular level and smaller, that energy as well as matter occurs in discrete units: When energy of an atom or molecule changes from one value to another, it does so in definite jumps, with no possible values in between. These quantum effects make phenomena on the atomic scale very different from what we are familiar with. When radiation encounters an atom, it can excite the atom to a higher internal energy level only if it can supply just the right amount of energy for the step. The reverse also occurs: When the energy level of an atom relaxes by a step, a discrete amount (quantum) of radiation energy is produced. The light emitted by a substance or absorbed by a substance can therefore serve to identify what the substance is, whether the substance is in the laboratory or is on the surface of a distant star.

Reactions in the nuclei of atoms involve far greater energy changes than reactions between the outer electron structures of atoms (that is, chemical reactions). When very heavy nuclei, such as those of uranium or plutonium, split into middle-weight ones, and when very light nuclei, such as those of hydrogen and helium, combine into somewhat heavier ones, large amounts of energy are released as radiation and rapidly moving particles. Fission of some heavy nuclei occurs spontaneously, producing extra neutrons that induce fission in more nuclei and thus giving rise to a chain reaction. The fusion of nuclei, however, occurs only if they collide at very great speeds (overcoming the electric repulsion between them), such as the collisions that occur at the very high temperatures produced inside a star or by a fission explosion.

THE MOTION OF THINGS

Motion is as much a part of the physical world as matter and energy are. Everything moves—atoms and molecules; the stars, planets, and moons; the earth and its surface and everything on its surface; all living things, and every part of living things. Nothing in the universe is at rest.

Since everything is moving, there is no fixed reference point against which the motion of things can be described. All motion is relative to whatever point or object we choose. Thus, a parked bus has no motion with reference to the earth's surface; but since the earth spins on its axis, the bus is moving about 1,000 miles per hour around the center of the earth. If the bus is moving down the highway, then a person walking up the aisle of the bus has one speed with reference to the bus, another with respect to the highway, and yet another with respect to the earth's center. There is no point in space that can serve as a reference for what is actually moving.

Changes in motion—speeding up, slowing down, changing direction—are due to the effects of forces. Any object maintains a constant speed and direction of motion unless an unbalanced outside force acts on it. When an unbalanced force does act on an object, the object's motion changes. Depending on the direction of the force relative to the direction of motion, the object may change its speed (a falling apple) or its direction of motion (the moon in its curved orbit), or both (a fly ball). The greater the amount of the unbalanced force, the more rapidly a given object's speed or direction of motion changes; the more massive an object is, the less rapidly its speed or direction changes in response to any given force. And whenever some thing A exerts a force on some thing B, B exerts an equally strong force back on A. For example, iron nail A pulls on magnet B with the same amount of force as magnet B pulls on iron nail A—but in the opposite direction. In most familiar situations, friction between surfaces brings forces into play that complicate the description of motion, although the basic principles still apply.

Some complicated motions can be described most conveniently not in terms of forces directly but in summary descriptions of the pattern of motion, such as vibrations and waves. Vibration involves parts of a system moving back and forth in much the same place, so the motion can be summarized by how frequently it is repeated and by how far a particle is displaced during a cycle. Another summary characteristic is the rate at which the vibration, when left to itself, dies down as its energy dissipates.

Vibrations may set up a traveling disturbance that spreads away from its source. Examples of such disturbances are sound, light, and earthquakes, which show some behavior very like that of familiar surface waves on water—changing direction at boundaries between media, diffracting around corners, and mutually interfering with one another in predictable ways. We therefore speak of sound waves, light waves, and so on, and the mathematics of wave behavior is useful in describing all these phenomena. Wave behavior can also be described in terms of how fast the disturbance propagates, and in terms of the distance between successive peaks of the disturbance (the wavelength).

The observed wavelength of a wave depends in part upon the relative motion of the source of the wave with respect to the observer. If the source is moving toward the observer (or vice versa), the wave is in effect compressed and perceived as shorter; if the source and observer are moving farther apart, the wave is in effect stretched out and perceived as longer. Both effects are evident in the apparent change in pitch of an automobile horn as it passes the observer. These apparent shifts in wavelength therefore provide information about relative motion. A particularly significant example of this shift is the change in the wavelength of light from stars and galaxies. Because the light emitted from most of them shifts toward longer wavelengths (that is, toward the red end of the spectrum), astronomers conclude that galaxies are all moving away from one another—and hence that we are in a generally expanding universe.

Wavelength can greatly influence how a wave interacts with matter—how well it is transmitted, absorbed, reflected, or diffracted. For example, the ways in which shock waves of different wavelengths travel through and reflect from layers of rock are an important clue as to what the interior of the earth is like. The interaction of electromagnetic waves with matter varies greatly with wavelength, both in how they are produced and in what their effects are. Different but somewhat overlapping ranges have been given distinctive names: radio waves, microwaves, radiant heat or infrared radiation, visible light, ultraviolet radiation, x rays, and gamma rays.

Materials that allow one range of wavelengths to pass through them may completely absorb others. For example, some gases in the atmosphere, including carbon dioxide and water vapor, are transparent to much of the incoming sunlight but not to the infrared radiation from the warmed surface of the earth, which bounces back into the atmosphere. Consequently, heat energy is trapped in the atmosphere. The temperature of the earth rises until its radiation output reaches equilibrium with the radiation input from the sun. Another atmospheric gas, ozone, absorbs some of the ultraviolet radiation in sunlight—the wavelengths that produce burning, tanning, and cancer in the skin of human beings.

Even within the named ranges of electromagnetic radiation, different wavelengths interact with matter in different ways. The most familiar example is that different wavelengths of visible light interact with our eyes differently, giving us the sensation of different colors. Things appear to have different colors because they reflect or scatter visible light of some wavelengths more than others, as in the case of plants that absorb blue and red wavelengths and reflect only green and yellow. When the atmosphere scatters sunlight—which is a mixture of all wavelengths—short-wavelength light (which gives us the sensation of blue) is scattered much more by air molecules than long-wavelength (red) light is. The atmosphere, therefore, appears blue and the sun seen through it by unscattered light appears reddened.

THE FORCES OF NATURE

The two forces we are commonly aware of are gravitational and electromagnetic. Everything in the universe exerts gravitational forces on everything else, although the effects are readily noticeable only when at least one very large mass is involved (such as a star or planet). Gravity is the force behind the fall of rain, the power of rivers, the pulse of tides; it pulls the matter of planets and stars toward their centers to form spheres, holds planets in orbit, and gathers cosmic dust together to form stars. Gravitational forces are thought of as involving a gravitational field that affects space around any mass. The strength of the field around an object is proportional to its mass and diminishes with distance from its center. For example, the earth's pull on an individual will depend on whether the person is, say, on the beach or far out in space.

The electromagnetic forces acting within and between atoms are trillions of trillions times stronger than the gravitational forces acting between them. On an atomic scale, electric forces between oppositely charged protons and electrons hold atoms and molecules together and thus are involved in all chemical reactions. On a larger scale, these forces hold solid and liquid materials together and act between objects when they are in contact (for example, the friction between a towel and a person's back, the impact of a bat on a ball). We usually do not notice the electrical nature of many familiar forces because the nearly equal densities of positive and negative electric charges in materials approximately neutralize each other's effects outside the material. But even a tiny imbalance in these opposite charges will produce phenomena that range from electric sparks and clinging clothes to lightning.

Depending on how many of the electric charges in them are free to move, materials show great differences in how much they respond to electric forces. At one extreme, an electrically insulating material such as glass or rubber does not ordinarily allow any passage of charges through it. At the other extreme, an electrically conducting material such as copper will offer very little resistance to the motion of charges, so electric forces acting on it readily produce a current of charges. (Most electrical wires are a combination of extremes: a very good conductor covered by a very good insulator.) In fact, at very low temperatures, certain materials can become superconductors, which offer zero resistance. In between low- and high-resistance materials are semiconducting materials in which the ease with which charges move may vary greatly with subtle changes in composition or conditions; these materials are used in transistors and computer chips to control electrical signals. Water usually contains charged molecular fragments of dissolved impurities that are mobile, and so it is a fairly good conductor.

Magnetic forces are very closely related to electric forces—the two can be thought of as different aspects of a single electromagnetic force. Both are thought of as acting by means of fields: an electric charge has an electric field in the space around it that affects other charges, and a magnet has a magnetic field around it that affects other magnets. What is more, moving electric charges produce magnetic fields and are affected by magnetic fields. This influence is the basis of many natural phenomena. For example, electric currents circulating in the earth's core give the earth an extensive magnetic field, which we detect from the orientation of our compass needles.

The interplay of electric and magnetic forces is also the basis of much technological design, such as electric motors (in which currents produce motion), generators (in which motion produces currents), and television tubes (in which a beam of moving electric charges is bent back and forth by a periodically changing magnetic field). More generally, a changing electric field induces a magnetic field, and vice versa.

Other types of forces operate only at the subatomic scale. For example, the nuclear force that holds particles together within the atomic nucleus is much stronger than the electric force, as is evident in the relatively great amounts of energy released by nuclear interactions.

Henri Rousseau, *Tropical Forest With Monkeys* (1910).

CHAPTER 5

THE LIVING ENVIRONMENT

People have long been curious about living things—how many different species there are, what they are like, where they live, how they relate to each other, and how they behave. Scientists seek to answer these questions and many more about the organisms that inhabit the earth. In particular, they try to develop the concepts, principles, and theories that enable people to understand the living environment better.

Living organisms are made of the same components as all other matter, involve the same kind of transformations of energy, and move using the same basic kinds of forces. Thus, all of the physical principles discussed in Chapter 4, "The Physical Setting," apply to life as well as to stars, raindrops, and television sets. But living organisms also have characteristics that can be understood best through the application of other principles.

This chapter offers recommendations on basic knowledge about how living things function and how they interact with one another and their environment. The chapter focuses on six major subjects: the diversity of life, as reflected in the biological characteristics of the earth's organisms; the transfer of heritable characteristics from one generation to the next; the structure and functioning of cells, the basic building blocks of all organisms; the interdependence of all organisms and their environment; the flow of matter and energy through the grand-scale cycles of life; and how biological evolution explains the similarity and diversity of life.

RECOMMENDATIONS

DIVERSITY OF LIFE

There are millions of different types of individual organisms that inhabit the earth at any one time—some very similar to each other, some very different. Biologists classify organisms into a hierarchy of groups and subgroups on the basis of similarities and differences in their structure and behavior. One of the most general distinctions among organisms is between plants, which get their energy directly from sunlight, and animals, which consume the energy-rich foods initially synthesized by plants. But not all organisms are clearly one or the other. For example, there are single-celled organisms without organized nuclei (bacteria) that are classified as a distinct group.

Animals and plants have a great variety of body plans, with different overall structures and arrangements of internal parts to perform the basic operations of making or finding food, deriving energy and materials from it, synthesizing new materials, and reproducing. When scientists classify organisms, they consider details of anatomy to be more relevant than behavior or general appearance. For example, because of such features as milk-producing glands and brain structure, whales and bats are classified as being more nearly alike than are whales and fish or bats and birds. At different degrees of relatedness, dogs are classified with fish as having backbones, with cows as having hair, and with cats as being meat eaters.

For sexually reproducing organisms, a species comprises all organisms that can mate with one another to produce fertile offspring. The definition of species is not precise, however; at the boundaries it may be difficult to decide on the exact classification of a particular organism. Indeed, classification systems are not part of nature. Rather, they are frameworks created by biologists for describing the vast diversity of organisms, suggesting relationships among living things, and framing research questions.

The variety of the earth's life forms is apparent not only from the study of anatomical and behavioral similarities and differences among organisms but also from the study of similarities and differences among their molecules. The most complex molecules built up in living organisms are chains of smaller molecules. The various kinds of small molecules are much the same in all life forms, but the specific sequences of components that make up the very complex molecules are characteristic of a given species. For example, DNA molecules are long chains linking just four kinds of smaller molecules, whose precise sequence encodes genetic information. The closeness or remoteness of the relationship between organisms can be inferred from the extent to which their DNA sequences are similar. The relatedness of organisms inferred from similarity in their molecular structure closely matches the classification based on anatomical similarities.

The preservation of a diversity of species is important to human beings. We depend on two food webs to obtain the energy and materials necessary for life. One starts with microscopic ocean plants and seaweed and includes animals that feed on them and animals that feed on those animals. The other one begins with land plants and includes animals that feed on them, and so forth. The elaborate interdependencies among species serve to stabilize these food webs. Minor disruptions in a particular location tend to lead to changes that eventually restore the system. But large disturbances of living populations or their environments may result in irreversible changes in the food webs. Maintaining diversity increases the likelihood that some varieties will have characteristics suitable to survival under changed conditions.

HEREDITY

One long-familiar observation is that offspring are very much like their parents but still show some variation: Offspring differ somewhat from their parents and from one another. Over many generations, these differences can accumulate, so organisms can be very different in appearance and behavior from their distant ancestors. For example, people have bred their domestic animals and plants to select desirable characteristics; the results are modern varieties of dogs, cats, cattle, fowl, fruits, and grains that are perceptibly different from their forebears. Changes have also been observed— in grains, for example—that are extensive enough to produce new species. In fact, some branches of descendants of the same parent species are so different from others that they can no longer breed with one another.

Instructions for development are passed from parents to offspring in thousands of discrete genes, each of which is now known to be a segment of a molecule of DNA. Offspring of asexual organisms (clones) inherit all of the parent's genes. In sexual reproduction of plants and animals, a specialized cell from a female fuses with a specialized cell from a male. Each of these sex cells contains an unpredictable half of the parent's genetic information. When a particular male cell fuses with a particular female cell during fertilization, they form a cell with one complete set of paired genetic information, a combination of one half-set from each parent. As the fertilized cell multiplies to form an embryo, and eventually a seed or mature individual, the combined sets are replicated in each new cell.

The sorting and combination of genes in sexual reproduction results in a great variety of gene combinations in the offspring of two parents. There are millions of different possible combinations of genes in the half apportioned into each separate sex cell, and there are also millions of possible combinations of the particular female and male sex cells.

However, new mixes of genes are not the only source of variation in the characteristics of organisms. Although genetic instructions may be passed down virtually unchanged for many thousands of generations, occasionally some of the information in a cell's DNA is altered. Deletions, insertions, or substitutions of DNA segments may occur spontaneously through random errors in copying, or may be induced by chemicals or radiation. If a mutated gene is in an organism's sex cell, copies of it may be passed down to offspring, becoming part of all their cells and perhaps giving the offspring new or modified characteristics. Some of these changed characteristics may turn out to increase the ability of the organisms that have it to thrive and reproduce, some may reduce that ability, and some may have no appreciable effect.

CELLS

All self-replicating life forms are composed of cells—from single-celled bacteria to elephants, with their trillions of cells. Although a few giant cells, such as hens' eggs, can be seen with the naked eye, most cells are microscopic. It is at the cell level that many of the basic functions of organisms are carried out: protein synthesis, extraction of energy from nutrients, replication, and so forth. The mechanisms by which these processes occur are similar in all living organisms. In addition, most cells perform certain specialized functions.

The main features of a cell are the membrane that surrounds the cell and controls what can enter and leave it; an internal fluid medium and protein skeleton that gives the cell shape and serves as a support medium for other internal parts of the cell; the nucleus, which contains the DNA; and many specialized organelles and other structures within the cell that are involved in the transport of materials, energy release, protein building, waste disposal, information feedback, and motility.

The genetic information encoded in DNA molecules provides instructions for assembling protein molecules. This code is virtually the same for all life forms. Thus, for example, if a gene from a human cell is placed in a bacterium, the chemical machinery of the bacterium will follow the gene's instructions and produce the same protein that would be produced in human

cells. Protein molecules are long, usually folded chains made from 20 different kinds of amino acid molecules. The function of each protein depends on its specific sequence of amino acids and the shape the chain takes as a consequence of attractions between the chain's parts. The mutation of a DNA segment of a gene may not make much difference, may fatally disrupt the operation of the cell, or may change the successful operation of the cell in a significant way (for example, it may foster uncontrolled replication, as in cancer). Thus, exposure of cells to certain chemicals and radiation tends to increase mutation rates and thereby increase chances of cancer.

The work of the cell is carried out mostly by the many different types of protein molecules it assembles. Some molecules assist in replicating genetic information, carrying out cell division, changing cell shape, repairing cell structures, and generally in catalyzing and regulating molecular interactions. Some of these molecules are exported from the cell: hormones, antibodies, digestive enzymes, carriers for oxygen and other molecules in the blood, and material for hair, nails, and other body structures.

Complex interactions among the myriad kinds of molecules in the cell may give rise to distinct cycles of activities, such as growth and division. Control of cell processes comes also from without: Cell behavior may be influenced by molecules from other parts of the organism or from other organisms (for example, hormones and neurotransmitters) that attach to or pass through the cell membrane and affect the rates of reaction among cell constituents.

In addition to the basic cellular functions common to all cells, most cells in multicelled organisms are specialized. They perform some special functions that others do not. For example, gland cells secrete hormones, muscle cells contract, and nerve cells conduct electrical signals. And yet all these cells are descendants of the single fertilized egg cell and have the same DNA information. As successive generations of cells form by division, small differences in their immediate environments cause them to develop slightly differently, by activating or inactivating different parts of the DNA information. Later generations of cells differ still further and eventually mature into cells as different as gland, muscle, and nerve cells.

INTERDEPENDENCE OF LIFE

Every species is linked, directly or indirectly, with a multitude of others in an ecosystem. Plants provide food, shelter, and nesting sites for other organisms. For their part, many plants depend upon animals for help in reproduction (bees pollinate flowers, for instance) and for certain nutrients (such as minerals in animal waste products). All animals are part of food webs that include plants and animals of other species (and sometimes the same species). The predator/prey relationship is common, with its offensive tools for predators—teeth, beaks, claws, venom, etc.—and its defensive tools for prey—camouflage to hide, speed to escape, shields or spines to ward off, irritating substances to repel. Some species come to depend very closely on others (for instance, pandas or koalas can eat only certain species of trees). Some species have become so adapted to each other that neither could survive without the other (for example, the wasps that nest only in figs and are the only insect that can pollinate them).

There are also other relationships between organisms. Parasites get nourishment from their host organisms, sometimes with bad consequences for the hosts. Scavengers and decomposers feed only on dead animals and plants. And some organisms have mutually beneficial relationships—for example, the bees that sip nectar from flowers and incidentally carry pollen from one flower to the next, or the bacteria that live in our intestines and incidentally synthesize some vitamins and protect the intestinal lining from germs.

But the interaction of living organisms does not take place on a passive environmental stage. Ecosystems are shaped by the nonliving environment of land and water— solar radiation, rainfall, mineral concentrations, temperature, and topography. The world contains a wide diversity of physical conditions, which creates a wide variety of environments: freshwater and oceanic, forest, desert, grassland, tundra, mountain, and many others. In all these environments, organisms use vital earth resources, each seeking its share in specific ways that are limited by other organisms. In every part of

the habitable environment, different organisms vie for food, space, light, heat, water, air, and shelter. The linked and fluctuating interactions of life forms and environment compose a total ecosystem; understanding any one part of it well requires knowledge of how that part interacts with the others.

The interdependence of organisms in an ecosystem often results in approximate stability over hundreds or thousands of years. As one species proliferates, it is held in check by one or more environmental factors: depletion of food or nesting sites, increased loss to predators, or invasion by parasites. If a natural disaster such as flood or fire occurs, the damaged ecosystem is likely to recover in a succession of stages that eventually results in a system similar to the original one.

Like many complex systems, ecosystems tend to show cyclic fluctuations around a state of approximate equilibrium. In the long run, however, ecosystems inevitably change when climate changes or when very different new species appear as a result of migration or evolution (or are introduced deliberately or inadvertently by humans).

FLOW OF MATTER AND ENERGY

However complex the workings of living organisms, they share with all other natural systems the same physical principles of the conservation and transformation of matter and energy. Over long spans of time, matter and energy cycle back and forth among living things, and between them and the physical environment. In these grand-scale cycles, the total amount of matter and energy remains constant, even though their form and location undergo continual change.

Almost all life on earth is ultimately maintained by transformations of energy from the sun. Plants capture the sun's energy and use it to synthesize complex, energy-rich molecules (chiefly sugars) from molecules of carbon dioxide and water. These synthesized molecules then serve, directly or indirectly, as the source of energy for the plants themselves and ultimately for all animals and decomposer organisms (such as bacteria and fungi). This is the food web: The organisms that consume the plants derive energy and materials from breaking down the plant molecules, use them to synthesize their own structures, and then are themselves consumed by other organisms. At each stage in the food web, some energy is stored in newly synthesized structures and some is dissipated into the environment as heat produced by the energy-releasing chemical processes in cells. A similar energy cycle begins with the capture of the sun's energy by minute marine organisms. Each successive stage in a food web captures only a small fraction of the energy content of organisms it feeds on.

The elements that make up the molecules of living things are continuously recycled. Chief among these elements are carbon, oxygen, hydrogen, nitrogen, sulfur, phosphorus, calcium, sodium, potassium, and iron. These and other elements, mostly occurring in energy-rich molecules, are passed along the food web and eventually are recycled by decomposers back to mineral nutrients usable by plants. Although there often may be local excesses and deficits, the situation over the whole earth is that organisms are dying and decaying at about the same rate as that at which new life is being synthesized. That is, the total living biomass stays roughly constant, there is a cyclic flow of materials from old to new life, and there is an irreversible flow of energy from captured sunlight into dissipated heat.

An important interruption in the usual flow of energy apparently occurred millions of years ago when the growth of land plants and marine organisms exceeded the ability of decomposers to recycle them. The accumulating layers of energy-rich organic material were gradually turned into great coal beds and oil pools by the pressure of the overlying earth. The energy stored in their molecular structure remains to us to release by combustion, and our modern civilization depends on immense amounts of energy from such fossil fuels recovered from the earth. By burning fossil fuels, we are finally passing most of the stored energy on to the environment as heat. We are also passing back to the atmosphere—in a relatively very short time—large amounts of carbon dioxide that had been removed from it slowly over millions of years.

The amount of life any environment can sustain is limited by its most basic resources: the inflow of energy, minerals, and water. Sustained productivity of an ecosystem requires sufficient energy for new products that are synthesized (such as trees and crops) and also for recycling completely the residue of the old (dead leaves, human sewage, etc.). When human technology intrudes, materials may accumulate as waste that is not recycled. When the inflow of resources is insufficient, the ecosystem draws upon its reserves of living and dead biomass. This process can result in accelerated soil leaching, desertification, or depletion of mineral reserves.

EVOLUTION OF LIFE

The earth's present-day life forms have evolved from common ancestors reaching back to the simplest one-cell organisms about three billion years ago. Modern ideas of evolution provide a scientific explanation for three main sets of observable facts about life on earth: (1) the enormous number of different life forms we see about us, (2) the systematic similarities in anatomy and molecular chemistry we see within that diversity, and (3) the sequence of changes in fossils found in successive layers of rock that have been formed over more than a billion years.

Since the beginning of the fossil record, many new life forms have appeared, and most old forms have disappeared. The many traceable sequences of changing anatomical forms, inferred from ages of rock layers, convince scientists that the accumulation of differences from one generation to the next has led eventually to species as different from one another as bacteria are from elephants. The molecular evidence substantiates the anatomical evidence from fossils and provides additional detail about the sequence in which various lines of descent branched off from one another.

Although details of the history of life on earth are still being pieced together from the combined geological, anatomical, and molecular evidence, the main features of that history are generally agreed upon. Life on earth has existed for three billion years. Prior to that, simple molecules may have formed complex organic molecules that very gradually formed into cells capable of self-replication. During the first two billion years of life, only microorganisms existed—

some of them apparently quite similar to bacteria and algae that exist today. With the development of cells with nuclei about a billion years ago, there was a great increase in the rate of evolution of increasingly complex, multicelled organisms. The rate of evolution of new species has been uneven since then, perhaps reflecting the varying rates of change in the physical environment.

A central concept of the theory of evolution is natural selection, which arises from three well-established observations: (1) There is some variation in heritable characteristics within every species of organism, (2) some of these characteristics will give individuals an advantage over others in surviving to maturity and reproducing, and (3) those individuals will be likely to have more offspring, which will themselves be more likely than others to survive and reproduce. The likely result is that over successive generations, the proportion of individuals that have inherited advantage-giving characteristics will tend to increase.

Selectable characteristics can include details of biochemistry, such as the molecular structure of hormones or digestive enzymes, and anatomical features that are ultimately produced in the development of the organism, such as bone size or fur length. They can also include more subtle features determined by anatomy, such as acuity of vision or pumping efficiency of the heart. By biochemical or anatomical means, selectable characteristics may also influence behavior, such as weaving a certain shape of web, preferring certain characteristics in a mate, or being disposed to care for offspring.

New heritable characteristics can result from new combinations of parents' genes or from mutations of them. Except for mutation of the DNA in an organism's sex cells, the characteristics that result from occurrences during the organism's lifetime cannot be biologically passed on to the next genera-

tion. Thus, for example, changes in an individual caused by use or disuse of a structure or function, or by changes in its environment, cannot be promulgated by natural selection.

By its very nature, natural selection is likely to lead to organisms with characteristics that are well adapted to survival in particular environments. Yet chance alone, especially in small populations, can result in the spread of inherited characteristics that have no inherent survival or reproductive advantage or disadvantage. Moreover, when an environment changes (in this sense, other organisms are also part of the environment), the advantage or disadvantage of characteristics can change. So natural selection does not necessarily result in long-term progress in a set direction. Evolution builds on what already exists, so the more variety that already exists, the more there can be.

The continuing operation of natural selection on new characteristics and in changing environments, over and over again for millions of years, has produced a succession of diverse new species. Evolution is not a ladder in which the lower forms are all replaced by superior forms, with humans finally emerging at the top as the most advanced species. Rather, it is like a bush: Many branches emerged long ago; some of those branches have died out; some have survived with apparently little or no change over time; and some have repeatedly branched, sometimes giving rise to more complex organisms.

The modern concept of evolution provides a unifying principle for understanding the history of life on earth, relationships among all living things, and the dependence of life on the physical environment. While it is still far from clear how evolution works in every detail, the concept is so well established that it provides a framework for organizing most of biological knowledge into a coherent picture.

Tao-chi, *Among Peaks and Pines (Mt. Huang)* (ca. 1701).

CHAPTER 6
THE HUMAN ORGANISM

As similar as we humans are in many ways to other species, we are unique among the earth's life forms in our ability to use language and thought. Having evolved a large and complex brain, our species has a facility to think, imagine, create, and learn from experience that far exceeds that of any other species. We have used this ability to create technologies and literary and artistic works on a vast scale, and to develop a scientific understanding of ourselves and the world.

We are also unique in our profound curiosity about ourselves: How are we put together physically? How were we formed? How do we relate biologically to other life forms and to our ancestors? How are we as individuals like or unlike other humans? How can we stay healthy? Much of the scientific endeavor focuses on such questions.

This chapter presents recommendations for what scientifically literate people should know about themselves as a species. Such knowledge provides a basis for increased awareness of both self and society. The chapter focuses on six major aspects of the human organism: human identity, the life cycle, the basic functions of the body, learning, physical health, and mental health. The recommendations on physical and mental health are included because they help relate the scientific understanding of the human organism to a major area of concern—personal well-being—common to all humans.

RECOMMENDATIONS

HUMAN IDENTITY

In most biological respects, humans are like other living organisms. For instance, they are made up of cells like those of other animals, have much the same chemical composition, have organ systems and physical characteristics like many others, reproduce in a similar way, carry the same kind of genetic information system, and are part of a food web.

Fossil and molecular evidence supports the belief that the human species, no less than others, evolved from other organisms. Evidence continues to accumulate and scientists continue to debate dates and lineage, but the broad outlines of the story are generally accepted. Primates—the classification of similar organisms that includes humans, monkeys and apes, and several other kinds of mammals—began to evolve from other mammals less than 100 million

years ago. Several humanlike primate species began appearing and branching about 5 million years ago, but all except one became extinct. The line that survived led to the modern human species.

Like other complex organisms, people vary in size and shape, skin color, body proportions, body hair, facial features, muscle strength, handedness, and so on. But these differences are minor compared to the internal similarity of all humans, as demonstrated by the fact that people from anywhere in the world can physically mix on the basis of reproduction, blood transfusions, and organ transplants. Humans are indeed a single species. Furthermore, as great as cultural differences between groups of people seem to be, their complex languages, technologies, and arts distinguish them from any other species.

Some other species organize themselves socially—mainly by taking on different specialized functions, such as defense, food collection, or reproduction—but they follow relatively fixed patterns that are limited by their genetic inheritance. Humans have a much greater range of social behavior—from playing card games to singing choral music, from mastering multiple languages to formulating laws.

One of the most important events in the history of the human species was the turn some 10,000 years ago from hunting and gathering to farming, which made possible rapid increases in population. During that early period of growth, the social inventiveness of the human species began to produce villages and then cities, new economic and political systems, recordkeeping—and organized warfare. Recently, the greater efficiency of agriculture and the control of infectious disease has further accelerated growth of the human population, which is now more than five billion.

Just as our species is biological, social, and cultural, so is it technological. Compared with other species, we are nothing special when it comes to speed, agility, strength, stamina, vision, hearing, or the ability to withstand extremes of environmental conditions. A variety of technologies, however, improves our ability to interact with the physical world. In a sense, our inventions have helped us make up for our biological disadvantages. Written records enable us to share and compile great amounts of information. Vehicles allow us to move more rapidly than other animals, to travel in many media (even in space), and to reach remote and inhospitable places. Tools provide us with very delicate control and with prodigious strength and speed. Telescopes, cameras, infrared sensors, microphones, and other instruments extend our visual, auditory, and tactile senses, and increase their sensitivity. Prosthetic devices and chemical and surgical intervention enable people with physical disabilities to function effectively in their environment.

LIFE CYCLE

A human develops from a single cell, formed by the fusion of an egg cell and a sperm cell; each contributes half of the cell's genetic information. Ovaries in females produce ripened egg cells, usually one per menstrual cycle; testes in males produce sperm cells in great numbers. Fertilization of an egg cell by a sperm ordinarily occurs after sperm cells are deposited near an egg cell. But fertilization does not always result: Sperm deposit may take place at the time of the female's menstrual cycle when no egg is present; contraceptive measures may be used to deliberately block or incapacitate egg or sperm; or one of the partners may be unable to produce viable sex cells.

Within a few hours of conception, the fertilized egg divides into two identical cells, each of which soon divides again, and so on, until there are enough to form a small sphere. Within a few days, this sphere embeds itself in the wall of the uterus, where the placenta nourishes the embryo by allowing the transfer of substances between the blood of the mother and that of the developing child. During the first three months of pregnancy, successive generations of cells organize into organs; during the second three months, all organs and body features develop; and during the last three months, further development and growth occur.

The developing embryo may be at risk as a consequence of its own genetic defects, the mother's poor health or inadequate diet during pregnancy, or her use of alcohol, tobacco, and other drugs. If an infant's development is incomplete when birth occurs, because of either premature birth or poor health care, the infant may not survive. After birth, infants may be at risk as a result of injury during birth or infection during or shortly after the event. The death rate of infants, therefore, varies greatly from place to place, depending on the quality of sanitation, hygiene, prenatal nutrition, and medical care. Even for infants who survive, poor prenatal conditions may lead to lower physical and mental capacities.

In normal children, mental development is characterized by the regular appearance of a set of abilities at successive stages. These include an enhancement of memory toward the end of the first month, speech sounds by the first birthday, connected speech by the second birthday, the ability to relate concepts and categories by the sixth birthday, and the ability to detect consistency or inconsistency in arguments by adolescence. The development of these increasingly more complex levels of intellectual competence is a function both of increasing brain maturity and of learning experiences. If appropriate kinds of stimulation are not available when the child is in an especially sensitive stage of development, some kinds of further biological and psychological development may be made more difficult or may even fail to occur.

This extraordinarily long period of human development—compared to that of other species—is related to the prominent role of the brain in human evolution. Most species are very limited in their repertory of behavior and depend for survival on predictable responses determined largely by genetic programming; mammals, and especially humans, depend far more on learned behavior. A prolonged childhood provides time and opportunities for the brain to develop into an effective instrument for intelligent living. This comes not only through play and interaction with older children and adults but also through exposure to the words and arts of people from other parts of the world and other times in history. The ability to learn persists throughout life and in some ways may improve as people build a base of ideas and come to understand how they learn best.

Developmental stages occur with somewhat different timing for different individuals, as a function of both differing physiological factors and differing experiences. Transition from one stage to another may be troublesome, particularly when biological changes are dramatic or when they are out of step with social abilities or others' expectations. Different societies place different meaning and importance on developmental stages and on the transitions from one to the next. For example, childhood is defined legally and socially as well as biologically, and its duration and meaning vary in different cultures and historical periods. In the United States, the onset of puberty—the maturation of the body in preparation for reproduction—occurs several years before an age generally considered physically and psychologically appropriate for parenthood and other adult functions.

Whether adults become parents, and (if they do) how many offspring they have, is determined by a wide variety of cultural and personal factors, as well as by biology. Technology has added greatly to the options available to people to control their reproduction. Chemical and mechanical means exist for preventing, detecting, or terminating pregnancies. Through such measures as hormone therapy and artificial insemination, it is also possible to bring about desired pregnancies that otherwise could not happen. The use of these technologies to prevent or facilitate pregnancy, however, is controversial and raises questions of social mores, ethics, religious belief, and even politics.

Aging is a normal—but still poorly understood—process in all humans. Its effects vary greatly among individuals. In general, muscles and joints tend to become less flexible, bones and muscles lose some mass, energy levels diminish, and the senses become less acute. For women, one major event in the aging process is menopause; sometime between the ages of 45 and 55, they undergo a major change in their production of sex hormones, with the result that they no longer have menstrual cycles and no longer release eggs.

The aging process in humans is associated not only with changes in the hormonal system but also with disease and injury, diet, mutations arising and accumulating in the cells, wear on tissues such as weight-bearing joints, psychological factors, and exposure to harmful substances. The slow accumulation of injurious agents such as deposits in arteries, damage to the lungs from smoking, and radiation damage to the skin, may produce noticeable disease. Sometimes diseases that appear late in life will affect brain function, including memory and personality. In addition, diminished physical capacity and loss of one's accustomed social role can result in anxiety or depression. On the other hand, many old people are able to get along quite well, living out independent and active lives, without prolonged periods of disability.

There appears to be a maximum life span for each species, including humans. Although some humans live more than a hundred years, most do not; the average length of life, including individuals who die in childhood, ranges from as low as 35 in some populations to as high as 75 in most industrialized nations. The high averages are due mostly to low death rates for infants and children but also to better sanitation, diet, and hygiene for most people, and to improved medical care for the old. Life expectancy also varies among different socioeconomic groups and by sex. The most common causes of death differ for various age, ethnic, and economic groups. In the United States, for example, fatal traffic accidents are most common among young males, heart disease causes more deaths in men than women, and infectious diseases and homicides cause more deaths among the poor than among the rich.

BASIC FUNCTIONS

The human body is a complex system of cells, most of which are grouped into organ systems that have specialized functions. These systems can best be understood in terms of the essential functions they serve: deriving energy from food, protection against injury, internal coordination, and reproduction. The continual need for energy engages the senses and skeletal muscles in obtaining food, the digestive system in breaking food down into usable compounds and in disposing of undigested food materials, the lungs in providing oxygen for combustion of food and discharging the carbon dioxide produced, the urinary system for disposing of other dissolved waste products of cell activity, the skin and lungs for getting rid of excess heat (into which most of the energy in food eventually degrades), and the circulatory system for moving all these substances to or from cells where they are needed or produced.

Like all organisms, humans have the means of protecting themselves. Self-protection involves using the senses in detecting danger, the hormone system in stimulating the heart and gaining access to emergency energy supplies, and the muscles in escape or defense. The skin provides a shield against harmful substances and organisms, such as bacteria and parasites. The immune system provides protection against the substances that do gain entrance into the body and against cancerous cells that develop spontaneously in the body. The nervous system plays an especially important role in survival; it makes possible the kind of learning humans need to cope with changes in their environment.

The internal control required for managing and coordinating these complex systems is carried out by the brain and nervous system in conjunction with the hormone-excreting glands. The electrical and chemical signals carried by nerves and hormones integrate the body as a whole. The many cross-influences between the hormones and nerves give rise to a system of coordinated cycles in almost all body functions. Nerves can excite some glands to excrete hormones, some hormones affect brain cells, the brain itself releases hormones that affect human behavior, and hormones are involved in transmitting signals between nerve cells. Certain drugs—legal and illegal—can affect the human body and brain by mimicking or blocking the hormones and neurotransmitters produced by the hormonal and nervous systems.

Reproduction ensures continuation of the species. The sexual urge is biologically driven, but how that drive is manifested among humans is determined by psychological and cultural factors. Sense organs and hormones are involved, as well as the internal and external sex organs themselves. The fact that sexual reproduction produces a greater genetic variation by mixing the genes of the parents plays a key role in evolution.

LEARNING

Among living organisms, much behavior is innate in the sense that any member of a species will predictably show certain behavior without having had any particular experiences that lead up to it (for example, a toad catching a fly that moves into its visual field). Some of this innate potential for behavior, however, requires that the individual develop in a fairly normal environment of stimuli and experience. In humans, for example, speech will develop in an infant without any special training if the infant can hear and imitate speech in its environment.

The more complex the brain of a species, the more flexible its behavioral repertoire is. Differences in the behavior of individuals arise partly from inherited predispositions and partly from differences in their experiences. There is continuing scientific study of the relative roles of inheritance and learning, but it is already clear that behavior results from the interaction of those roles, not just a simple sum of the two. The apparently unique human ability to transmit ideas and practices from one generation to the next, and to invent new ones, has resulted in the virtually unlimited variations in ideas and behavior that are associated with different cultures.

Learning muscle skills occurs mostly through practice. If a person uses the same muscles again and again in much the same way (throwing a ball), the pattern of movement may become automatic and no longer require any conscious attention. The level of skill eventually attained depends on an individual's innate abilities, on the amount of practice, and on the feedback of information and reward. With enough practice, long sequences of behaviors can become virtually automatic (driving a car along a familiar route, for instance). In this case, a person does not have to concentrate on the details of coordinating sight and muscle movements and can also engage in, say, conversation at the same time. In an emergency, full attention can rapidly be focused back on the unusual demands of the task.

Learning usually begins with the sensory systems through which people receive information about their bodies and the physical and social world around them. The way each person perceives or experiences this information depends not only on the stimulus itself but also on the physical context in which the stimulus occurs and on numerous physical, psychological, and social factors in the beholder. The senses do not give people a mirror image of the world but respond selectively to a certain range of stimuli. (The eye, for example, is sensitive to only a small fraction of the electromagnetic spectrum.) Furthermore, the senses selectively filter and code information, giving some stimuli more importance, as when a sleeping parent hears a crying baby, and others less importance, as when a person adapts to and no longer notices an unpleasant odor. Experiences, expectations, motivations, and emotional levels can all affect perceptions.

Much of learning appears to occur by association: If two inputs arrive at the brain at approximately the same time, they are likely to become linked in memory, and one perception will lead to an expectation of the other. Actions as well as perceptions can be associated. At the simplest possible level,

behavior that is accompanied or followed by pleasant sensations is likely to occur again, whereas behavior followed by unpleasant sensations is less likely to occur again. Behavior that has pleasant or unpleasant consequences only under special conditions will become more or less likely when those special conditions occur. The strength of learning usually depends on how close the inputs are matched in time and on how often they occur together. However, there can be some subtle effects. For example, a single, highly unpleasant event following a particular behavior may result in the behavior being avoided ever after. On the other hand, rewarding a particular behavior even only every now and then may result in very persistent behavior.

But much of learning is not so mechanical. People tend to learn much from deliberate imitation of others. Nor is all learning merely adding new information or behaviors. Associations are learned not only among perceptions and actions but also among abstract representations of them in memory—that is, among ideas. Human thinking involves the interaction of ideas, and ideas about ideas, and thus can produce many associations internally without further sensory input.

People's ideas can affect learning by changing how they interpret new perceptions and ideas: People are inclined to respond to, or seek, information that supports the ideas they already have and on the other hand to overlook or ignore information that is inconsistent with the ideas. If the conflicting information is not overlooked or ignored, it may provoke a reorganization of thinking that makes sense of the new information, as well as of all previous information. Successive reorganizations of one part or another of people's ideas usually result from being confronted by new information or circumstances. Such reorganization is essential to the process of human maturation and can continue throughout life.

PHYSICAL HEALTH

To stay in good operating condition, the human body requires a variety of foods and experiences. The amount of food energy (calories) a person requires varies with body size, age, sex, activity level, and metabolic rate. Beyond just energy, normal body operation requires substances to add to or replace the materials of which it is made: unsaturated fats, trace amounts of a dozen elements whose atoms play key roles, and some traces of substances that human cells cannot synthesize—including some amino acids and vitamins. The normal condition of most body systems requires that they perform their adaptive function: For example, muscles must effect movement, bones must bear loads, and the heart must pump blood efficiently. Regular exercise, therefore, is important for maintaining a healthy heart/ lung system, for maintaining muscle tone, and for keeping bones from becoming brittle.

Good health also depends on the avoidance of excessive exposure to substances that interfere with the body's operation. Chief among those that each individual can control are tobacco (implicated in lung cancer, emphysema, and heart disease), addictive drugs (implicated in psychic disorientation and nervous-system disorders), and excessive amounts of alcohol (which has negative effects on the liver, brain, and heart). In addition, the environment may contain dangerous levels of substances (such as lead, some pesticides, and radioactive isotopes) that can be harmful to humans. Therefore, the good health of individuals also depends on people's collective effort to monitor the air, soil, and water and to take steps to keep them safe.

Other organisms also can interfere with the human body's normal operation. Some kinds of bacteria or fungi may infect the body to form colonies in preferred organs or tissues. Viruses invade healthy cells and cause them to synthesize more viruses, usually killing those cells in the process. Infectious disease also may be caused by animal parasites, which may take up residence in the intestines, bloodstream, or tissues.

The body's own first line of defense against infectious agents is to keep them from entering or settling in the body. Protective mechanisms include skin to block them, tears and saliva to carry them out, and stomach and vaginal secretions to kill them. Related means of protecting against invasive organisms include keeping the skin clean, eating properly, avoiding contaminated foods and liquids, and generally avoiding needless exposure to disease.

The body's next line of defense is the immune system. White blood cells act both to surround invaders and to produce specific antibodies that will attack them (or facilitate attack by other white cells). If the individual survives the invasion, some of these antibodies remain—along with the capability of quickly producing many more. For years afterward, or even a lifetime, the immune system will be ready for that type of organism and be able to limit or prevent the disease. A person can "catch a cold" many times because there are many varieties of germs that cause similar symptoms. Allergic reactions are caused by unusually strong immune responses to some environmental substances, such as those found in pollen, on animal hair, or in certain foods. Sometimes the human immune system can malfunction and attack even healthy cells. Some viral diseases, such as AIDS, destroy critical cells of the immune system, leaving the body helpless in dealing with multiple infectious agents and cancerous cells.

Infectious diseases are not the only threat to human health, however. Body parts or systems may develop impaired function for entirely internal reasons. Some faulty operations of body processes are known to be caused by deviant genes. They may have a direct, obvious effect, such as causing easy bleeding, or they may only increase the body's susceptibility to developing particular diseases, such as clogged arteries or mental depression. Such genes may be inherited, or they may result from mutation in one cell or a few cells during an individual's own development. Because one properly functioning gene of a pair may be sufficient to perform the gene's function, many genetic diseases do not appear unless a faulty form of the gene is inherited from both parents (who, for the same reason, may have had no symptoms of the disease themselves).

The fact that most people now live in physical and social settings that are very different from those to which human physiology was adapted long ago is a factor in determining the health of the population in general. One modern "abnormality" in industrialized countries is diet, which once included chiefly raw plant and animal materials but now includes excess amounts of refined sugar, saturated fat, and salt, as well as caffeine, alcohol, nicotine, and other drugs. Lack of exercise is another change from the much more active life-style of prehistory. There are also environmental pollutants and the psychological stress of living in a crowded, hectic, and rapidly changing social environment. On the other hand, new medical techniques, efficient health care delivery systems, improved sanitation, and a fuller public understanding of the nature of disease give today's humans a better chance of staying healthy than their forebears had.

MENTAL HEALTH

Good mental health involves the interaction of psychological, biological, physiological, social, and cultural systems. It is generally regarded as the ability to cope with the ordinary circumstances people encounter in their personal, professional, and social lives. Ideas about what constitutes good mental health vary, however, from one culture to another and from one time period to another. Behavior that may be regarded as outright insanity in one culture may be regarded in another as merely eccentricity or even as divine inspiration. In some cultures, people may be classified as mentally ill if they persistently express disagreement with religious or political authorities. Ideas about what constitutes proper treatment for abnormal mental states differ also. Evidence of abnormal thinking that would be deliberately punished in one culture may be treated in other cultures by social involvement, by isolation, by increased social support, by prayers, by extensive interviews, or by medical procedures.

Individuals differ greatly in their ability to cope with stressful environments. Stresses in childhood may be particularly difficult to deal with, and, because they may shape the subsequent experience and thinking of the child, they may have long-lasting effects on a person's psychological health and social adjustment. And people also differ in how well they can cope with psychological disturbance when it occurs. Often, people react to mental distress by denying that they have a psychological problem. Even when people recognize that they do have such a problem, they may not have the money, time, or social support necessary to seek help. Prolonged disturbance of behavior may result in strong reactions from families, work supervisors, and civic authorities that add to the stress on the individual.

Diagnosis and treatment of mental disturbances can be particularly difficult because much of people's mental life is not usually accessible even to them. When we remember someone's name, for example, the name just seems to come to us—the conscious mind has no idea of what the search process was. Similarly, we may experience anger or fear or depression without knowing why. According to some theories of mental disturbance, such feelings may result from exceptionally upsetting thoughts or memories that are blocked from becoming conscious. In treatment based on such theories, clues about troubling unconscious thoughts may be sought in the patient's dreams or slips of the tongue, and the patient is encouraged to talk long and freely to get the ideas out in the open where they can be dealt with.

Some kinds of severe psychological disturbance once thought to be purely spiritual or mental have a basis in biological abnormality. Destruction of brain tissue by tumors or broken blood vessels can produce a variety of behavioral symptoms, depending on which locations in the brain are affected. For example, brain injuries may affect the ability to put words together comprehensibly or to understand the speech of others, or may cause meaningless emotional outbursts. Deficiency or excess of some chemicals produced in the brain may result in hallucinations and chronic depression. The mental deterioration that sometimes occurs in the aged may be caused by actual disease of the brain. Biological abnormality does not necessarily produce the psychological malfunction by itself, but it may make individuals exceptionally vulnerable to other causes of disturbance.

Conversely, intense emotional states have some distinct biochemical effects. Fear and anger cause hormones to be released into the bloodstream that prepare the body for action—fight or flight. Psychological distress may also affect an individual's vulnerability to biological disease. There is some evidence that intense or chronic emotional states can sometimes produce changes in the nervous, visceral, and immune systems. For example, fear, anger, depression, or even just disappointment may lead to the development of headaches, ulcers, and infections. Such effects can make the individual even more vulnerable to psychological stress—creating a vicious circle of malfunction. On the other hand, there is evidence that social contacts and support may improve an individual's ability to resist certain diseases or may minimize their effects.

George Catlin, *A Little Sioux Village* (1857/1869).

CHAPTER 7

HUMAN SOCIETY

As a species, we are social beings who live out our lives in the company of other humans. We organize ourselves into various kinds of social groupings, such as nomadic bands, villages, cities, and countries, in which we work, trade, play, reproduce, and interact in many other ways. Unlike other species, we combine socialization with deliberate changes in social behavior and organization over time. Consequently, the patterns of human society differ from place to place and era to era and across cultures, making the social world a very complex and dynamic environment.

Insight into human behavior comes from many sources. The views presented here are based principally on scientific investigation, but it should also be recognized that literature, drama, history, philosophy, and other nonscientific disciplines contribute significantly to our understanding of ourselves. Social scientists study human behavior from a variety of cultural, political, economic, and psychological perspectives, using both qualitative and quantitative approaches. They look for consistent patterns of individual and social behavior and for scientific explanations of those patterns. In some cases, such patterns may seem obvious once they are pointed out, although they may not have been part of how most people consciously thought about the world. In other cases, the patterns—as revealed by scientific investigation—may show people that their long-held beliefs about certain aspects of human behavior are incorrect.

This chapter covers recommendations about human society in terms of individual and group behavior, social organizations, and the processes of social change. It is based on a particular approach to the subject: the sketching of a comprehensible picture of the world that is consistent with the findings of the separate disciplines within the social sciences—such as anthropology, economics, political science, sociology, and psychology—but without attempting to describe the findings themselves or the underlying methodologies.

The chapter describes seven key aspects of human society: cultural effects on human behavior, the organization and behavior of groups, the processes of social change, social trade-offs, forms of political and economic organization, mechanisms for resolving conflict among groups and individuals, and national and international social systems. Although many of the ideas are relevant to all human societies, this chapter focuses chiefly on the social characteristics of the present-day United States.

RECOMMENDATIONS

CULTURAL EFFECTS ON BEHAVIOR

Human behavior is affected both by genetic inheritance and by experience. The ways in which people develop are shaped by social experience and circumstances within the context of their inherited genetic potential. The scientific question is just how experience and hereditary potential interact in producing human behavior.

Each person is born into a social and cultural setting—family, community, social class, language, religion—and eventually develops many social connections. The characteristics of a child's social setting affect how he or she learns to think and behave, by means of instruction, rewards and punishment, and example. This setting includes home, school, neighborhood, and also, perhaps, local religious and law enforcement agencies. Then there are also the child's mostly informal interactions with friends, other peers, relatives, and the entertainment and news media. How individuals will respond to all these influences, or even which influence will be the most potent, tends not to be predictable. There is, however, some substantial similarity in how individuals respond to the same pattern of influences—that is, to being raised in the same culture. Furthermore, culturally induced behavior patterns, such as speech patterns, body language, and forms of humor, become so deeply imbedded in the human mind that they often operate without the individuals themselves being fully aware of them.

Every culture includes a somewhat different web of patterns and meanings: ways of earning a living, systems of trade and government, social roles, religions, traditions in clothing and foods and arts, expectations for behavior, attitudes toward other cultures, and beliefs and values about all of these activities. Within a large society, there may be many groups, with distinctly different subcultures associated with region, ethnic origin, or social class. If a single culture is dominant in a large region, its values may be considered correct and may be promoted—not only by families and religious groups but also by schools and governments. Some subcultures may arise among special social categories (such as

teenagers, business executives, and criminals), some of which may cross national boundaries (such as rock musicians and scientists).

Fair or unfair, desirable or undesirable, social distinctions are a salient part of almost every culture. The form of the distinctions varies with place and time, sometimes including rigid castes, sometimes tribal or clan hierarchies, sometimes a more flexible social class. Class distinctions are made chiefly on the basis of wealth, education, and occupation, but they are also likely to be associated with other subcultural differences, such as dress, dialect, and attitudes toward school and work. These economic, political, and cultural distinctions are recognized by almost all members of a society—and resented by some of them.

The class into which people are born usually determines what language, diet, tastes, and interests they will have as children, and it strongly influences how they will perceive the social world. Moreover, class affects what pressures and opportunities they will experience and therefore what paths their lives are likely to take—including schooling, occupation, marriage, use of leisure time, and acquisition of wealth.

The ease with which someone can move up to another class, though, varies greatly with time and place. Throughout most of human history, people have been almost certain to live and die in the class into which they were born. The times of greatest upward mobility have occurred when a society has been undertaking new enterprises (for example, in territory or technology) and thus has needed more people in higher-class occupations. In some parts of the world today, increasing numbers of people are escaping from poverty through economic or educational opportunity, while in other parts, increasing numbers are being impoverished.

What is considered to be acceptable human behavior varies from culture to culture and from time period to time period. Every social group has generally accepted ranges of behavior for its members, with perhaps some specific standards for subgroups, such as adults and children, females and males, artists and athletes. Unusual behaviors may be considered either merely amusing, or distasteful, or punishably criminal. Some normal behavior in one culture may be considered unacceptable in another. For example, aggressively competitive behavior is considered rude in highly cooperative cultures. Conversely, in some subcultures of a highly competitive society, such as that of the United States, a lack of interest in competition may be regarded as being out of step. Although the world has a wide diversity of cultural traditions, there are some kinds of behavior (such as incest, violence against kin, theft, and rape) that are considered unacceptable in almost all of them.

The social consequences considered appropriate for unacceptable behavior also vary widely between, and even within, different societies. Punishment of criminals ranges from fines or humiliation to imprisonment or exile, from beatings or mutilation to execution. The form of appropriate punishment is affected by theories of its purpose to prevent or deter the individual from repeating the crime, or to deter others from committing the crime, or simply to cause suffering for its own sake in retribution. The success of punishment in deterring crime is difficult to study, in part because of ethical limitations on experiments assigning different punishments to similar criminals, and in part because of the difficulty of holding other factors constant.

Technology has long played a major role in human behavior. The high value placed on new technological invention in many parts of the world has led to increasingly rapid and inexpensive communication and travel, which in turn has led to the rapid spread of fashions and ideas in clothing, food, music, and forms of recreation. Books, magazines, radio, and television describe ways to dress, raise children, make money, find happiness, get married, cook, and make love. They also implicitly promote values, aspirations, and priorities by the way they portray the behavior of people such as children, parents, teachers, politicians, and athletes, and the attitudes they display toward violence, sex, minorities, the roles of men and women, and lawfulness.

GROUP ORGANIZATION AND BEHAVIOR

In addition to belonging to the social and cultural settings into which they are born, people voluntarily join groups based on shared occupations, beliefs, or interests (such as unions, political parties, or clubs). Membership in these groups influences how people think of themselves and how others think of them. These groups impose expectations and rules that make the behavior of members more predictable and that enable each group to function smoothly and retain its identity. The rules may be informal and conveyed by example, such as how to behave at a social gathering, or they may be written rules that are strictly enforced. Formal groups often signal the kind of behavior they favor by means of rewards (such as praise, prizes, or privileges) and punishments (such as threats, fines, or rejection).

Affiliation with any social group, whether one joins it voluntarily or is born into it, brings some advantages of larger numbers: the potential for pooling resources (such as money or labor), concerted effort (such as strikes, boycotts, or voting), and identity and recognition (such as organizations, emblems, or attention from the media). Within each group, the members' attitudes, which often include an image of their group as being superior to others, help ensure cohesion within the group but can also lead to serious conflict with other groups. Attitudes toward other groups are likely to involve stereotyping—treating all members of a group as though they were the same and perceiving in those people's actual behavior only those qualities that fit the observer's preconceptions—which leads to social discrimination. Blind respect for professionals such as doctors or clergymen, as well as blind disrespect for people who are, say, minorities or women, are some of the forms of discrimination that arise from stereotyping.

The behavior of groups cannot be understood solely as the aggregate behavior of individuals. It is not possible, for example, to understand modern warfare by summing up the aggressive tendencies of individuals. A person may behave very differently in a crowd—say, when at a football game, at a religious service, or on a picket line—than when alone or with family members. Several children together may vandalize a building, even though none of them would do it on his or her own. By the same token, an adult will often be more generous and responsive to the needs of others as a member of, say, a club or religious group than he or she would be inclined to be in private. The group situation provides the rewards of companionship and acceptance for going along with the shared action of the group and makes it difficult to assign blame or credit to any one person.

Social organizations may serve many purposes beyond those for which they formally exist. Private clubs that exist ostensibly for recreation are frequently important places for engaging in business transactions; universities that formally exist to promote learning and scholarship may help to promote or to reduce class distinctions; and business and religious organizations often have political and social agendas that go beyond making a profit or ministering to people. In many cases, an unstated purpose of groups is to exclude people in particular categories from their activities—yet another form of discrimination.

SOCIAL CHANGE

Societies, like species, evolve in directions that are opened or constrained in part by internal forces such as technological developments or political traditions. The conditions of one generation limit and shape the range of possibilities open to the next. On the one hand, each new generation learns the society's cultural forms and thus does not have to reinvent strategies for producing food, handling conflict, educating young people, governing, and so forth. It also learns aspirations for how society can be maintained and improved. On the other hand, each new generation must address unresolved problems from the generation before: tensions that may lead to war, wide-scale drug abuse, poverty and deprivation, racism, and a multitude of private and group grievances. Slavery in the early history of the United States, for example, still has serious consequences for black Americans and for the U.S. economy, education, health care, and judicial systems in general. Grievances may be relieved just enough to make people tolerate them, or they may overflow into revolution against the struc-

ture of the society itself. Many societies continue to perpetuate centuries-old disputes with others over boundaries, religion, and deeply felt beliefs about past wrongs.

Governments generally attempt to engineer social change by means of policies, laws, incentives, or coercion. Sometimes these efforts work effectively and actually make it possible to avoid social conflict. At other times they may precipitate conflict. For example, setting up agricultural communes in the Soviet Union against the farmers' wishes to farm their own private land was achieved only with armed force and the loss of millions of lives. The liberation of slaves in the United States came only as one consequence of a bloody civil war; a hundred years later, elimination of explicit racial segregation was achieved in some places only by use of legislative action, court injunctions, and armed military guard—and continues to be a major social issue.

External factors—including war, migration, colonial domination, imported ideas, technology or plagues, and natural disasters—also shape the ways in which each society evolves. The outlook of the Soviet Union, for example, is strongly influenced by the devastating losses it suffered in both world wars. The societies of American Indians were ravaged and displaced by the diseases and warfare brought by colonists from Europe. In the United States, forcible importation of Africans and successive waves of immigrants from Europe, Latin America, and Asia have greatly affected the political, economic, and social systems (such as labor, voting blocs, and educational programs), as well as adding to the nation's cultural variety. Natural disasters such as storms or drought can cause failure of crops, bringing hardship and famine, and sometimes migration or revolution.

Convenient communication and transportation also stimulate social change. Groups previously isolated geographically or politically become ever more aware of different ways of thinking, living, and behaving, and sometimes of the existence of vastly different standards of living. Migrations and mass media lead not only to cultural mixing but also to the extinction of some cultures and the rapid evolution of others. Easy worldwide communications and transportation force confrontations of values and expectations—sometimes deliberately as propaganda, sometimes just incidentally, as in the pursuit of commercial interests.

The size of the human population, its concentration in particular places, and its pattern of growth are influenced by the physical setting and by many aspects of culture: economics, politics, technology, history, and religion. In response to economic concerns, national governments set very different policies—some to reduce population growth, some to increase it. Some religious groups also take a strong stand on population issues. Leaders of the Roman Catholic church, for example, have long campaigned against birth control, whereas, in recent years, religious leaders of other major faiths have endorsed the use of birth control to restrict family size.

Quite apart from government policy or religious doctrine, many people decide whether to have a child on the basis of practical matters such as the health risk to the mother, the value or cost of a child in economic or social terms, the amount of living space, or a personal feeling of suitability as parents. In some parts of the world—and in poorly educated groups in any country—couples have little knowledge of, or access to, modern-birth control information and technology. In the United States, the trend toward casual adolescent sexual relations has led to increasing numbers of unexpected and unwanted pregnancies.

In turn, social systems are influenced by population—its size, its rate of change, and its proportions of people with different characteristics (such as age, sex, and language). Great increase in the size of a population requires greater job specialization, new government responsibilities, new kinds of institutions, and the need to marshal a more complex distribution of resources. Population patterns, particularly when they are changing, are also influential in changing social priorities. The greater the variety of subcultures, the more diverse the provisions that have to be made for them. As the size of a social group increases, so may its influence on society. The influence may be through markets (such as young people who, as a group, buy more athletic equipment), voting power (for example, old people are less likely to vote for school bond legislation), or recognition of need by social planners (for example, more mothers who work outside the home will require child-care programs).

SOCIAL TRADE-OFFS

Choices among alternative benefits and costs are unavoidable for individuals or for groups. To gain something we want or need, it is usually necessary to give up something we already have, or at least give up an opportunity to have gained something else instead. For example, the more the public spends as a whole on government-funded projects such as highways and schools, the less it can spend on defense (if it has already decided not to increase revenue or debt).

Social trade-offs are not always economic or material. Sometimes they arise from choices between our private rights and the public good: laws concerning cigarette smoking in public places, cleaning up after pets, and highway speed limits, for instance, restrict the individual freedom of some people for the benefit of others. Or choices may arise between esthetics and utility. For example, a proposed large-scale apartment complex may be welcomed by prospective tenants but opposed by people who already live in the neighborhood.

Different people have different ideas of how trade-offs should be made, which can result in compromise or in continuing discord. How different interests are served often depends on the relative amounts of resources or power held by individuals or groups. Peaceful efforts at social change are most successful when the affected people are included in the planning, when information is available from all relevant experts, and when the values and power struggles are clearly understood and incorporated into the decision-making process.

There is often a question of whether a current arrangement should be improved or whether an entirely new arrangement should be invented. On the one hand, repeatedly patching up a troublesome situation may make it just tolerable enough that the large-scale change of the underlying problem is never undertaken. On the other hand, rushing to replace every system that has problems may create more problems than it solves.

It is difficult to compare the potential benefits of social alternatives. One reason is that there is no common measure for different forms of good—for example, no measure by which wealth and social justice can be compared directly. Another reason is that different groups of people place greatly differing values on even the same kind of social good—for example, on public education or the minimum wage. In a very large population, value comparisons are further complicated by the fact that a very small percentage of the population can be a large number of people. For example, in a total population of 100 million, a rise in the unemployment rate of only one-hundredth of 1 percent (which some people would consider trivially small) would mean a loss of 10,000 jobs (which other people would consider very serious).

Judgments of consequences in social trade-offs tend to involve other issues as well. One is a distance effect: The farther away in distance or the further away in time the consequences of a decision are, the less importance we are likely to give them. City dwellers, for instance, are less likely to support national crop-support legislation than are farmers, and farmers may not wish to have their federal tax dollars pay for inner-city housing projects. As individuals, we find it difficult to resist an immediate pleasure even if the long-term consequences are likely to be negative, or to endure an immediate discomfort for an eventual benefit. As a society, similarly, we are likely to attach more importance to immediate benefits (such as rapidly using up our oil and mineral deposits) than to long-term consequences (shortages that we or our descendants may suffer later).

The effect of distance in judging social trade-offs is often augmented by uncertainty about whether potential costs and benefits will occur at all. Sometimes it is possible to estimate the probabilities of several possible outcomes of a social decision—for example, that sexual intercourse without birth control results in pregnancy about one time out of four. If relative value measures can also be placed on all the possible outcomes, the probabilities and value measures can be combined to estimate which alternative would be the best bet. But even when both probabilities and value measures are available, there may be debate about how to put the information together. People may be so afraid of some particular risk, for example, that they insist that it be reduced to as close to zero as possible, regardless of what other benefits or risks are involved.

And finally, decisions about social alternatives are usually complicated by the fact that people are reactive. When a social program is undertaken to achieve some intended effect, the inventiveness of people in promoting or resisting that effect will always add to the uncertainty of the outcome.

FORMS OF POLITICAL AND ECONOMIC ORGANIZATION

In most of the world's countries, national power and authority are allocated to various individuals and groups through politics, usually by means of compromises between conflicting interests. Through politics, governments are elected or appointed, or, in some cases, created by armed force. Governments have the power to make, interpret, and enforce the rules and decisions that determine how countries are run.

The rules that governments make encompass a wide range of human affairs, including commerce, education, marriage, medical care, employment, military service, religion, travel, scientific research, and the exchange of ideas. A national government—or, in some cases, a state or local government—is usually given responsibility for services that individuals or private organizations are believed not to be able to perform well themselves. The U.S. Constitution, for example, requires the federal government to perform only a few such functions: the delivery of mail, the taking of the census, the minting of money, and military defense. However, the increasing size and complexity of U.S. society has led to a vast expansion of government activities.

Today, the federal government is directly involved in such areas as education, welfare, civil rights, scientific research, weather prediction, transportation, preservation of national resources such as national parks, and much more. Decisions about the responsibilities that national, state, and local governments should have are negotiated among government officials, who are influenced by their constituencies and by centers of power such as corporations, the military, agricultural interests, and labor unions.

The political and economic systems of nations differ in many ways, including the means of pricing goods and services; the sources of capital for new ventures; government-regulated limits on profits; the collecting, spending, and controlling of money; and the relationships of managers and workers to each other and to government. The political system of a nation is closely intertwined with its economic system, refereeing the economic activity of individuals and groups at every level.

It is useful to think of the economy of a nation as tending toward one or the other of two major theoretical models. At one theoretical extreme is the purely capitalist system, which assumes that free competition produces the best allocation of resources, the greatest productivity and efficiency, and the lowest costs. Decisions about who does what and who gets what are made naturally as consumers and businesses interact in the marketplace, where prices are strongly influenced by how much something costs to make or do and how much people are willing to pay for it. Most enterprises are initiated by individuals or voluntary groups of people. When more resources are needed than are available to any one person (such as to build a factory), they may be obtained from other people, either by taking out loans from banks or by selling ownership shares of the business to other people. High personal motivation to compete requires private ownership of productive resources (such as land, factories, and ships) and minimal government interference with production or trade. According to capitalist theory, individual initiative, talent, and hard work are rewarded with success and wealth, and individual political and economic rights are protected.

At the other theoretical extreme is the purely socialist system, which assumes that the wisest and fairest allocation of resources is achieved through government planning of what is produced and who gets it at what cost. Most enterprises are initiated and financed by the government. All resources of production are owned by the state, on the assumption that private ownership causes greed and leads to the exploitation of workers by owners. According to socialist theory, people contribute their work and tal-

ents to society not for personal gain but for the social good; and the government provides benefits for people fairly, on the basis of their relative needs, not their talent and effort. The welfare of the society as a whole is regarded as being more important than the rights of any individuals.

There are, however, no nations with economic systems at either the capitalist or the socialist extreme; rather, the world's countries have at least some elements of both. Such a mixture is understandable in practical terms.

In a purely capitalist system, on the one hand, competition is seldom free because for any one resource, product, or service, a few large corporations or unions tend to monopolize the market and charge more than open competition would allow. Discrimination based on economically irrelevant social attitudes (for example, against minorities and women, in favor of friends and relatives) further distorts the ideal of free competition. And even if the system is efficient, it tends to make some individuals very rich and some very poor. Thus, the United States, for example, tries to limit the extreme effects of its basically capitalist economic system by mean of selective government intervention in the free-market system. This intervention includes tax rates that increase with wealth; unemployment insurance; health insurance; welfare support for the poor; laws that limit the economic power of any one corporation; regulation of trade among the states; government restrictions on unfair advertising, unsafe products, and discriminatory employment; and government subsidization of agriculture and industry.

On the other hand, a purely socialist economy, even though it may be more equitable, tends toward inefficiency by neglecting individual initiative and by trying to plan every detail of the entire national economy. Without some advantages in benefits to motivate people's efforts, productivity tends to be low. And without individuals having the freedom to make decisions on their own, short-term variations in supply and demand are difficult to respond to. Moreover, underground economies spring up to match realities of supply and demand for consumer products. Therefore, many socialist systems allow some measure of open competition and acknowledge the importance of individual initiative and ownership. Most economies throughout the world today are undergoing change—some adopting more capitalist policies and practices, and others adopting more socialist ones.

SOCIAL CONFLICT

There is conflict in all human societies, and all societies have systems for regulating it. Conflict between people or groups often arises from competition for resources, power, and status. Family members compete for attention. Individuals compete for jobs and wealth. Nations compete for territory and prestige. Different interest groups compete for influence and the power to make rules. Often the competition is not for resources but for ideas—one person or group wants to have the ideas or behavior of another group suppressed, punished, or declared illegal.

Social change can be potent in evoking conflict. Rarely if ever is a proposed social, economic, or political change likely to benefit every component of a social system equally, and so the groups that see themselves as possible losers resist. Mutual animosities and suspicions are aggravated by the inability of both proponents and opponents of any change to predict convincingly what all of the effects will be of making the change or of not making it. Conflict is particularly acute when only a few alternatives exist with no compromise possible—for example, between surrender and war or between candidate A and candidate B. Even though the issues may be complex and people may not be initially very far apart in their perceptions, the need to decide one way or the other can drive people into extreme positions to support their decision as to which alternative is preferable.

In family groups and small societies, laws are laid down by recognized authorities, such as parents or elders. All groups—from university faculties to local scout troops—have formalized procedures for making rules and arbitrating disputes. On a larger scale, government provides mechanisms for dealing with conflict by making laws and administering them. In a democracy, the political system arbitrates social conflict by means of elections. Candidates for office advertise their intentions to make and modify rules, and people vote for whoever

they believe has the best combination of intentions and the best chances of effectively carrying them out. But the need to make complex social trade-offs tends to prevent politicians from accomplishing all of their intentions when in office.

The desire for complete freedom to come and go as one pleases, carry weapons, and organize demonstrations may conflict with a desire for public security. The desire for resolute, efficient decision making—in the extreme, a dictatorship—may conflict with a desire for public participation—in the extreme, a democracy in which everyone votes on everything. The creation of laws and policies typically involves elaborate compromises negotiated among diverse interest groups. Small groups of people with special interests that they consider very important may be able to persuade their members to vote on the basis of that single issue and thereby demand concessions from a more diffuse majority.

Even when the majority of the people in a society agree on a social decision, the minority who disagree may have some protection. In the U.S. political system, for example, federal and state governments have constitutions that establish rights for citizens that cannot be changed by elected officials no matter how large a majority supports those officials. Changes in those constitutions usually require super majorities, of two-thirds or three-quarters of all voters, rather than just greater than one-half. A similar protection of political rights is provided by the two-house system in the federal legislature and in most state legislatures. In Congress, for instance, the lower house has representation in proportion to population, so that every citizen in the country is equally represented. However, the upper house has exactly two members from every state, regardless of its population—thereby ensuring that the citizens of any state, however tiny, have the same representation as those of any other state, however large.

In addition, societies have developed many informal ways of airing conflict, including debates, strikes, demonstrations, polls, advertisements, and even plays, songs, and cartoons. The mass media provide the free means for (and may even encourage) small groups of people with a grievance to make highly visible public statements. Any of these ways and means may either release tensions and promote compromise or inflame and further polarize differences. The failure to resolve or to moderate conflicts leads to tremendous stress on the social system. Inability or unwillingness to change may result in a higher level of conflict: lawsuits, sabotage, violence, or full-scale revolutions or wars.

Intergroup conflict, lawful or otherwise, does not necessarily end when one segment of society finally manages to effect a decision in its favor. The resisting groups may then launch efforts to reverse, modify, or circumvent the change, and so the conflict persists. Conflict can, however, also solidify group action; both nations and families tend to be more unified during times of crisis. Sometimes group leaders use this knowledge deliberately to provoke conflict with an outside group, thus reducing tensions and consolidating support within their own group.

WORLDWIDE SOCIAL SYSTEMS

Nations and cultures are increasingly dependent on one another through international economic systems and shared environmental problems (such as the global effects of nuclear warfare, deforestation, and acid rain). They also learn more about one another through international travel and use of mass media. More and more, the global system is becoming a tightly knit web in which a change in any one part of the web has significant effects on the others. For instance, local conflicts spread beyond their borders to involve other nations; fluctuating oil supplies affect economic productivity, trade balances, interest rates, and employment throughout the world. The wealth, security, and general welfare of almost all nations are interrelated. There is a growing consensus among the leaders of most nations that isolationist policies are no longer sustainable and that such global issues as controlling the spread of nuclear weapons and protecting the world's monetary system from wild fluctuations can be accomplished only by all nations acting in concert.

Nations interact through a wide variety of formal and informal arrangements. Formal ones include diplomatic relations, military and economic alliances, and global organizations such as the United Nations and the World Bank. Unlike national governments, however, global organizations often have only limited authority over their members. Other arrangements include cultural exchanges, the flow of tourists, student exchanges, international trade, and the activities of nongovernment organizations with worldwide membership (such as Amnesty International, anti-hunger campaigns, the Red Cross, and sports organizations).

The wealth of a nation depends on the effort and skills of its workers, its natural resources, and the capital and technology available to it for making the most of those skills and resources. Yet national wealth depends not only on how much a nation can produce for itself but also on the balance between how much its products are sought by other nations and how much of other nations' products it seeks. International trade does not result just from countries lacking certain resources or products, such as oil or various food crops or efficient automobiles. Even if a country can produce everything it needs by itself, it still benefits from trade with other countries. If a country does the things it does most efficiently (in terms of either quality or cost, or both) and sells its products to other nations, such a system theoretically enables all participating countries to come out ahead.

There are, however, many practical influences that distort the economic reality of international trade. For instance, such trade may be thwarted by fear of exploitation by economically or politically more powerful nations, by the desire to protect special groups of workers who would lose out to foreign economic competition, and by the unwillingness to become dependent on foreign countries for certain products that could become unavailable in the case of future conflicts.

Because of increasing international ties, the distinctions between international policy and domestic policy may be unclear in many cases. For example, policies that determine what kinds of cars or clothes we buy, and at what prices, are based on foreign trade and an international balance of payments. Agricultural production at home depends on foreign markets as well as domestic policies. Even though international markets may be to the advantage of all countries, they may be greatly to the disadvantage of particular groups of people within countries. The cheap production of cars in Asian countries, for example, may benefit car buyers all over the world but may also put automakers in other countries out of work. Domestic policies may thus be needed to avoid hardship for such groups; those policies in turn will affect international trade. Nations with strong internal consensus on their own religious or political ideologies may pursue foreign policies that aggressively promote the spread of such ideologies in other countries and undermine groups with competing ideas.

The growing interdependence of world social, economic, and ecological systems makes it difficult to predict the consequences of social decisions. Changes anywhere in the world can have amplified effects elsewhere, with increased benefits to some people and increased costs to others. There is also the possibility of some changes producing instability and uncertainty that are to the disadvantage of all. Worldwide stability may depend on nations establishing more reliable systems of doing business and exchanging information, developing monitoring mechanisms to warn of global catastrophes (such as famine and nuclear war), and reducing the large gap in the standard of living between the richest and the poorest nations. Nations, like all participants in social systems, are already finding it to their advantage to suffer some short-term losses to achieve the longer-term benefits of a stable world economy. Trading off national interests against global ones, however, may turn out to be the most important social decision of all.

Georgia O'Keeffe, *City Night* (1926).

CHAPTER 8

THE DESIGNED WORLD

The world we live in has been shaped in many important ways by human action. We have created technological options to prevent, eliminate, or lessen threats to life and the environment and to fulfill social needs. We have dammed rivers and cleared forests, made new materials and machines, covered vast areas with cities and highways, and decided—sometimes willy-nilly—the fate of many other living things.

In a sense, then, many parts of our world are designed—shaped and controlled, largely through the use of technology—in light of what we take our interests to be. We have brought the earth to a point where our future well-being will depend heavily on how we develop and use and restrict technology. In turn, that will depend heavily on how well we understand the workings of technology and the social, cultural, economic, and ecological systems within which we live.

This chapter sets forth recommendations about certain key aspects of technology, with emphasis on the major human activities that have shaped our environment and lives. The chapter starts with a discussion of the effects of the human presence on the globe, and then focuses on eight basic technology areas: agriculture, materials, manufacturing, energy sources, energy use, communication, information processing, and health technology.

RECOMMENDATIONS

THE HUMAN PRESENCE

The earth's population has already doubled three times during the past century. Even at that, the human presence, which is evident almost everywhere on the earth, has had a greater impact than sheer numbers alone would indicate. We have developed the capacity to dominate most plant and animal species—far more than any other species can—and the ability to shape the future rather than merely respond to it.

Use of that capacity has both advantages and disadvantages. On the one hand, developments in technology have brought enormous benefits to almost all people. Most people today have access to goods and services that were once luxuries enjoyed only by the wealthy—in transportation, communication, nutrition, sanitation, health care, entertainment, and so on. On the other hand, the very behavior that made it possible for the human species to prosper so

rapidly has put us and the earth's other living organisms at new kinds of risk. The growth of agricultural technology has made possible a very large population but has put enormous strain on the soil and water systems that are needed to continue sufficient production. Our antibiotics cure bacterial infection, but may continue to work only if we invent new ones faster than resistant bacterial strains emerge.

Our access to and use of vast stores of fossil fuels have made us dependent on a nonrenewable resource. In our present numbers, we will not be able to sustain our way of living on the energy that current technology provides, and alternative technologies may be inadequate or may present unacceptable hazards. Our vast mining and manufacturing efforts produce our goods, but they also dangerously pollute our rivers and oceans, soil, and atmosphere. Already, by-products of industrialization in the atmosphere may be depleting the ozone layer, which screens the planet's surface from harmful ultraviolet rays, and may be creating a buildup of carbon dioxide, which traps heat and could raise the planet's average temperatures significantly. The environmental consequences of a nuclear war, among its other disasters, could alter crucial aspects of all life on earth.

From the standpoint of other species, the human presence has reduced the amount of the earth's surface available to them by clearing large areas of vegetation; has interfered with their food sources; has changed their habitats by changing the temperature and chemical composition of large parts of the world environment; has destabilized their ecosystems by introducing foreign species, deliberately or accidentally; has reduced the number of living species; and in some instances has actually altered the characteristics of certain plants and animals by selective breeding and more recently by genetic engineering.

What the future holds for life on earth, barring some immense natural catastrophe, will be determined largely by the human species. The same intelligence that got us where we are—improving many aspects of human existence and introducing new risks into the world—is also our main resource for survival.

AGRICULTURE

Throughout history, most people have had to spend a great deal of their time getting food and fuel. People began as nomadic hunters and gatherers, using as food the animals and plants they found in the environment. Gradually, they learned how to expand their food supplies by using processing technology (such as pounding, salting, cooking, and fermenting). And they also learned how to use some usually inedible parts of animals and plants to make such things as tools, clothes, and containers. After many thousands of years of hunting and gathering, the human species developed ways of manipulating plants and animals to provide better food supplies and thereby support larger populations. People planted crops in one place and encouraged growth by cultivating, weeding, irrigating, and fertilizing. They captured and tamed animals

for food and materials and also trained them for such tasks as plowing and carrying loads; later, they raised such animals in captivity.

More advances in agriculture came over time as people learned not only to use but also to modify life forms. At first, they could control breeding only by choosing which of their animals and plants would reproduce. Combinations of the natural variety of characteristics could thus be attempted, to improve the domesticity, hardiness, and productivity of plant and animal species. To preserve the great variety of naturally adapted crop species that are available for crossbreeding, seed banks are set up around the world; their importance is evident in the international negotiations about who has what rights to those genetic resources.

In the twentieth century, the success of modern genetics has helped to increase the natural variability within plant species by using radiation to induce mutations, so that there are more choices for selective breeding. Scientists are now learning how to modify the genetic material of organisms directly. As we learn more about how the genetic code works (it is virtually the same for all life forms), it is becoming possible to move genes from one organism to another. With knowledge of what genetic code sequences control what functions, some characteristics can be transferred from one species to another; this technique may eventually lead to the design of new characteristics. For example, plants can be given the genetic program for synthesizing substances that give them resistance to insect predators.

One factor in improved agricultural productivity in recent decades has been the control of plant and animal pests. In the United States in the past, and elsewhere in the world still, a large fraction of farm products was lost to weeds, rodents, insects, and disease-causing microorganisms. The widespread use of insecticides, herbicides, and fungicides has greatly increased useful farm output. There are problems, however. One is that pesticides may also act harmfully on other organisms in the environment, sometimes far from where they are used, and sometimes greatly concentrated by water runoff and the food web. Insecticides used to control the boll weevil, for instance, killed off its natural predators, making the weevil problem worse. Another problem is that the effectiveness of the pesticides may diminish as organisms develop genetically determined resistance to them, thereby requiring increased amounts of pesticides or the development of new ones.

Consequently, more environmentally harmonious use of technology is being explored. This work involves the careful design and use of chemicals and a more knowledgeable diversification of crops, changing the crops planted on a particular tract of land from crops that deplete some constituent of the soil to crops that replenish it. Changing crops can also reduce the likelihood that particular crop diseases will get a foothold. An alternative to the chemical control of pests is introducing organisms from other ecosystems in an effort to reduce the number of pests in the agricultural ecosystem (such as by using foreign insects that feed on local weeds). This approach also carries some risk of an introduced organism's becoming a pest itself.

Agricultural productivity has grown through the use of machines and fertilizers. Machines and the fossil-fuel engines needed to power them have made it possible for one person to cultivate and harvest more land, to cultivate more different kinds of land, and to feed and use the parts and products of greater numbers of plants and animals. Fertilizers are widely used in the western hemisphere to supplement inadequate soil nutrients, in place of the manure used in many other parts of the world. One risk in the heavy use of machinery and fertilizers is the temptation to exhaust the soil from overuse. For that reason, the U.S. government encourages agricultural producers to take land out of production periodically and to take steps to restore the natural richness of the soil.

For many centuries, most food was consumed or marketed within a few dozen miles of where it was grown. Technology has revolutionized agricultural markets through transportation and communication. The many improvements in land productivity have led to availability of far more food in some areas than is needed for the local population. The development of rapid and cheap transportation reduces spoilage of food, as do treatment, additives, refrigera-

tion, and packaging. But rapid, long-distance distribution of farm products also requires rapid, long-distance distribution means for selling and routing them. Together, modern transportation and communication systems enable food to be marketed and consumed thousands of miles from where it is produced. These long-distance connections, however, do entail vulnerability to disruption.

When most markets were local, bad weather could cause great ups and downs in well-being for farmers and consumers. Now, because food is distributed through a world-wide market, consumers in wealthy nations have much less worry about an inadequate food supply. On the other hand, bad weather anywhere in the world can affect markets elsewhere. The concern of government for maintaining the national food supply for consumers and for protecting farmers from disastrous ups and downs in income has led to many forms of control of agriculture, which include controlling how land is used, what products are sold, and at what prices.

Only a century ago, a majority of workers in the United States were engaged in farming. Now, because technology has so greatly increased the efficiency of agriculture, only a tiny proportion (only about 2 percent) of the population is directly involved in production. There are, however, many more people involved in producing agricultural equipment and chemicals, and in the processing, storage, transportation, and distribution of food and fiber. The rapid reduction in the number of farmers needed to produce the nation's food has caused great shifts of population out of rural communities, resulting in the virtual disappearance of what was only recently the predominant way of life.

MATERIALS

Technology is based on the use and application of a great variety of materials, some of which occur naturally, some of which are produced by means of mixing or treating, and some of which are synthesized from basic materials. All materials have certain physical properties, such as strength, density, hardness, flexibility, durability, imperviousness to water and fire, and ease of conducting a flow of heat or electric current. These properties determine the use to which the materials are put by manufacturers, engineers, and others involved in technology.

For much of human history, materials technology was based chiefly on the use of natural materials such as plants, animal products, and minerals. Over time, people learned that the characteristics of natural materials could be changed by processing, such as the tanning of leather and the firing of clay. Later, they discovered that materials could be physically combined—mixed, layered, or bonded together—to get combinations of the characteristics of several different materials (for example, different kinds of wood laminated in a bow, steel rods embedded in concrete, zinc plated onto steel, and fibers interwoven in cloth). They also learned that the fine control of processes such as the tempering of steel or the annealing of glass could significantly improve some properties.

Since the 1960s, materials technology has focused increasingly on the synthesis of materials with entirely new properties. This process usually involves mixing substances together, as has been done for thousands of years with metal alloys. Typically, though, chemical changes are involved, and the properties of the new material may be entirely different from those of its constituents. Some new materials, such as plastics, are synthesized in chemical reactions that link long chains of atoms together. Plastics can be designed to have a wide variety of properties for different uses, from automobile and space vehicle parts, to food packaging and fabrics, to artificial hip joints and dissolving stitches. Ceramics, too, can be designed to have a variety of properties, and they can even be made such that the properties differ greatly from one ceramic to another (for example, the extremely low electrical conductivity of ceramic insulators, the controllable conductivity of ceramic semiconductors, and the virtually infinite conductivity of ceramic superconductors). Some materials can even be designed to adapt to various environments—such as all-weather oil and variable-density sunglasses.

The growth of technology has led us to use some materials from the environment much more rapidly than they can be replaced by natural processes. Forests in many countries have been greatly reduced

during the past few hundred years, and ore deposits are being depleted. There is a continuing search for substitute materials—and in many cases they have been found or invented.

Increasingly, the disposal of used materials has become a problem. Some used materials, such as organic wastes, can be returned safely to the environment—although as the population grows, the task becomes more difficult and more expensive. But some materials, such as plastics, are not easily recycled; nor do they degrade quickly when returned to the environment. Still other used materials—radioactive waste being the most dramatic but not the only example—are so hazardous for such a long time that how best to dispose of them is not clear and is the subject of widespread controversy. Solving these problems of disposal will require systematic efforts that include both social and technological innovations.

MANUFACTURING

Making things requires a great variety of tools. The growth of technology in general has been greatly helped by improvement in the fineness and sharpness of cutting tools, the force that can be applied, the temperature at which heat can be concentrated, the swiftness with which operations can occur, and the consistency with which operations can be repeated. Such tools are an essential factor in modern manufacturing, which is based largely on the need to produce great numbers of products of uniform quality (such as automobiles and wristwatches) and much smaller numbers of products of extremely high quality (such as space vehicles and atomic clocks).

Modern manufacturing processes usually involve three major steps: (1) obtaining and preparing raw materials; (2) mechanical processing such as shaping, joining, and assembling; and (3) coating, testing, inspecting, and packaging. In all of these steps, there are choices for how to sequence tasks and how to perform them, so the organization of tasks to optimize productivity is another major component of manufacturing.

Modern factories tend to specialize in making specific products. When a large number of nearly identical things are made on a continuous basis at the same place, it is possible to make them much more cheaply than if they were made separately. Such cost-effectiveness is achieved by bringing workers together with machines, energy sources, and raw materials or component parts. The maintenance and repair of products are also likely to be easier when production is centralized because parts can be made that are interchangeable between units and even between different models.

Production is increasingly automated. In some settings, robots are used to perform the repetitive tasks of mass production. Instructions for processing are used to control the processes electronically, rather than having to be interpreted and carried out by people. The flexibility in control makes it possible to design and use multipurpose machines that can customize a line of products. Such machines may also enable manufacturers to introduce a new line of products without first making a special set of new machines.

The design of manufacturing systems, whether automated or not, can be highly complex. First, the sequence of operations admits of many possibilities, from which highly efficient and cost-effective ones must be selected. Then, for any chosen sequence, a great number of flows of materials and timings of operations must be controlled, monitored, and coordinated. Many subtleties of human skill and judgment may be difficult to specify precisely; often, experts are not able to explain exactly what they do or how they do it. Control by computers makes efficient operation of highly complex manufacturing systems possible, but it still requires human supervision to deal with the unforeseen or unforeseeable.

The evolution of production has changed the nature of work. In the past, a craft worker could work at the same tasks for a lifetime with little change in product or technique. Large-scale production in one place led to an extreme of specialization: each worker doing just one simple task over and over again, rather than putting together complete products. Increasing automation requires less direct labor and fewer skilled crafts, but more engineering, computer programming, quality control, supervision, and maintenance. Although it may reduce workers' feelings of boredom and unimportance that result from endlessly repeating

the same minor tasks, automation also reduces the workers' control and may eliminate some workers' jobs even while it creates others. Flexibility and skill in learning a succession of new job roles have become increasingly important as the pace of technological change quickens.

ENERGY SOURCES

Industry, transportation, urban development, agriculture, and most other human activities are closely tied to the amount and kind of energy available. Energy is required for technological processes: taking apart, putting together, moving around, communicating, and getting raw materials, and then working them and recycling them.

Different sources of energy and ways of using them have different costs, implications, and risks. Some of the resources—direct sunlight, wind, and water—will continue to be available indefinitely. Plant fuels—wood and grasses—are self-renewing, but only at a limited rate and only if we plant as much as we harvest. Fuels already accumulated in the earth—coal, oil and natural gas, and uranium—will become more difficult to obtain as the most readily available sources run out. When scarcity threatens, new technology may make it possible to use the remaining sources better by digging deeper, processing lower-concentration ores, or mining the ocean bed. Just when they will run out completely, however, is difficult to predict. The ultimate limit may be prohibitive cost rather than complete disappearance—a question of when the energy required to obtain the resources becomes greater than the energy those resources will provide.

Sunlight is the ultimate source of most of the energy we use. It becomes available to us in several ways: The energy of sunlight is captured directly in plants, and it heats the air, land, and water to cause wind and rain. But the flux of energy is fairly weak, and large collection systems are necessary to concentrate energy for most technological uses: Hydroelectric energy technology uses rainwater concentrated in rivers by runoff from vast land areas; windmills use the flow of air produced by the heating of large land and ocean surfaces; and electricity generated from wind power and directly from sunlight falling on light-sensitive surfaces requires very large collection systems. Small-scale energy production for household use can be achieved in part by using windmills and direct solar heating, but cost-efficient technology for the large-scale use of windmills and solar heating has not yet been developed.

For much of history, burning wood was the most common source of intense energy for cooking, for heating dwellings, and for running machines. Most of the energy used today is derived from burning fossil fuels, which have stored sunlight energy that plants collected over millions of years. Coal was the most widely used fossil fuel until recently. But in the last century, oil and its associated natural gas have become preferred because of their ease of collection, multiple uses in industry, and ability to be concentrated into a readily portable source of energy for vehicles such as cars, trucks, trains, and airplanes. All burning of fossil fuels, unfortunately, dumps into the atmosphere waste products that may threaten health and life; the mining of coal underground is extremely hazardous to the health and safety of miners, and can leave the earth scarred; and oil spills can endanger marine life. Returning to the burning of wood is not a satisfactory alternative, for that too adds so-called greenhouse gases to the atmosphere; and overcutting trees for fuel depletes the forests needed to maintain healthy ecosystems both locally and worldwide.

But there are other sources of energy. One is the fission of the nuclei of heavy elements, which—compared to the burning of fossil fuels—releases an immense quantity of energy in relation to the mass of material used. In nuclear reactors, the energy generated is used mostly to boil water into steam, which drives electric generators. The required uranium is in large, although ultimately limited, supply. The waste products of fission, however, are highly radioactive and remain so for thousands of years. The technical problem of reasonably safe disposal of these fission products is compounded by public fear of radioactivity and worry about the sabotage of nuclear power plants and the theft of nuclear materials to make weapons. Controlled nuclear fusion reactions are a potentially much greater source of energy, but the technology has not yet proved feasible. Fusion reactions would use fuel materials that are safer in themselves, although there would still be a problem of disposing of worn-out construction materials made radioactive by the process. And as always with new technology, there may be some unanticipated risks.

ENERGY USE

Energy must be distributed from its source to where it is to be used. For much of human history, energy had to be used on site—at the windmill or water mill, or close to the forest. In time, improvement in transportation made it possible for fossil fuels to be burned far from where they were mined, and intensive manufacturing could spread much more widely. In this century, it has been common to use energy sources to generate electricity, which can deliver energy almost instantly along wires far from the source. Electricity, moreover, can conveniently be transformed into and from other kinds of energy.

As important as the amount of energy available is its quality: the extent to which it can be concentrated and the convenience with which it can be used. A central factor in technological change has been how hot a fire could be made. The discovery of new fuels, the design of better ovens and furnaces, and the forced delivery of air or pure oxygen have progressively increased the temperature available for firing clay and glass, smelting metal ores, and purifying and working metals. Lasers are a new tool for focusing radiation energy with great intensity and control, and they are being developed for a growing number of applications—from making computer chips and performing eye surgery to communicating by satellite.

During any useful transformation of energy from one form to another, there is inevitably some dissipation of energy into the environment. Except for the energy bound in the structure of manufactured materials, most of our uses of energy result in all of it eventually dissipating away, slightly warming the environment and ultimately radiating into space. In this practical sense, energy gets "used up," even though it is still around somewhere.

People have invented ingenious ways of deliberately bringing about energy transformations that are useful to them. These ways range from the simple acts of throwing rocks (which transforms biochemical energy into motion) and starting fires (chemical energy into heat and light), to using such complex devices as steam engines (heat energy into motion), electric generators (motion into electrical energy), nuclear fission reactors (nuclear energy into heat), and solar converters (radiation energy into electrical energy). In the operation of these devices, as in all phenomena, the useful energy output—that is, what is available for further change—is always less than the energy input, with the difference usually appearing as heat. One goal in the design of such devices is to make them as efficient as possible—that is, to maximize the useful output for a given input.

Consistent with the general differences in the global distribution of wealth and development, energy is used at highly unequal rates in different parts of the world. Industrialized nations use tremendous amounts of energy for chemical and mechanical processes in factories, creating synthetic materials, producing fertilizer for agriculture, powering industrial and personal transportation, heating and cooling buildings, lighting, and communications. The demand for energy at a still greater rate is likely as the world's population grows and more nations industrialize. Along with large-scale use, there is large-scale waste (for example, vehicles with more power than their function warrants and buildings insufficiently insulated against heat transfer). But other factors, especially an increase in the efficiency of energy use, can help reduce the demand for additional energy.

Depletion of energy sources can be slowed by both technical and social means. Technical means include maximizing the usefulness that we realize from a given input of energy by means of good design of the transformation device, by means of insulation where we want to restrict heat flow (for example, insulating hot-water tanks), or by doing something with the heat as it leaks out. Social means include government, which may restrict low-priority uses of energy or may establish requirements for efficiency (such as in automobile engines) or for insulation (as in house construction). Individuals also may make energy efficiency a consideration in their own choice and use of technology (for example, turning out lights and driving high-efficiency cars)—either to conserve energy as a matter of principle or to reduce their personal long-term expenses. As always, there are trade-offs. For example, better-insulated houses restrict ventilation and thus may increase the indoor accumulation of pollutants.

COMMUNICATION

People communicate frequently, if not always well. Hundreds of different languages have evolved to fit the needs of the people who use them. Because languages vary widely in sound, structure, and vocabulary and because language is so culturally bound, it is not always easy to translate from one to another with precision. Written communications—from personal letters to books and junk mail—crisscross continents and reach the farthest outposts. Telephones, radio, television, satellites, sound and optical recordings, and other forms of electronic communication have increased the options and added to the flow of information.

Communication involves a means of representing information, a means of transmitting and receiving it, and some assurance of fidelity between what is sent and what is received. Representation requires coding information in some transmission medium. In human history, the natural media have been mechanical contact (touch), chemicals (smell), sound waves (speech and hearing), and visible light (vision). But reliability and permanence require a medium for recording information. The reliable medium that developed first was the marking of solid materials—wood, clay, stone, and eventually paper. Today, we also mark, microscopically, plastic disks and magnetic tape. These modified materials can endure for many years, and can be moved great distances with their encoded information intact.

With the invention of devices to generate and control electric current, information could be encoded as changes in current and could be conveyed over long distances by wire almost instantaneously. With the discovery of radio waves, the same information could be encoded as changes in wave pattern and distributed in all directions through the atmosphere without the need of connecting wires. Particularly important was the invention of electronic amplifiers, in which a weak electrical signal controls the flow of a much stronger electric current, impressing it with the same pattern of information. Recently, the efficient control of light waves in lasers has made possible the encoding and transmitting of information as pulses in light intensity over optical fibers.

Information can be coded in analog or digital form. For example, originally both wired and wireless electric communication were only in the digital form of off-and-on bursts, requiring an artificial code to represent letters and numbers. A great advantage came with the invention of electronics— devices to transform sound and light signals into electrical signals, and vice versa. Electronics made it possible to transmit analog signals that represent subtle variations in sound or light and to transcribe those signals as continuous variations in some medium. The ability to transcribe information microscopically and to transmit information at very high rates now makes possible the reduction of distortion and noise in processing analog signals by returning to the reliability of off-and-on digital signals. Analog signals of all sorts can now be sampled and represented as numbers, stored or transmitted in that form, and conveniently processed by computers, and perhaps returned to analog form for sound or graphic display.

The basic technical challenge of communication is to keep the signal large compared to the noise, which always tends to increase when information is recorded, transformed, or transmitted. The ratio of signal to noise can be improved by boosting the signal or by reducing the noise. Signals can be kept strong by amplification or by preventing energy loss (as by focusing them in a narrow beam of waves). Noise can be limited by isolating the signal from external noise sources (as by shielding microphone cables) or by reducing internal sources of noise (as by cooling an amplifier). A very different way to minimize errors from noise in communication is by means of repetition or some other form of redundancy that allows comparison and detection of errors. Some redundancy is always desirable in communication, because otherwise a single error may completely change the meaning of a message.

Communication sometimes requires security. Mail can be intercepted and copied, telephone wires can be tapped, over-the-air communications can be monitored. Privacy can be protected by preventing access to signals (for example, by using locks and passwords) or by preventing interpretation of them (such as by using secret codes). The creation of secret codes that are extremely difficult to figure out is an interesting application of number theory in mathematics. As the techniques of providing security improve, however, so do techniques for penetrating it.

INFORMATION PROCESSING

Technology has long played an important role in collecting, storing, and retrieving information, as well as in transporting it. The invention of writing, tables of data, diagrams, mathematical formulas, and filing systems have all increased the amount of information we can handle and the speed with which we can process it. Large amounts of information are essential for the operation of modern societies; indeed, the generation, processing, and transfer of information is becoming the most common occupation of workers in industrialized countries.

Information is most useful when it is organized and represented by orderly collections of symbols. People use tables, indexes, alphabetical lists, and hierarchical networks to organize large amounts of data. The best way to store information depends on what is to be done with it. Information stored with one purpose in mind may be very troublesome to retrieve for other purposes (for example, the alphabetical listing of telephone numbers is ideal if one knows a person's name, but not if one knows only the address). Multipurpose data bases enable the information to be located in several different ways (for example, books listed by author or title or subject). A typical feature of such information systems is attaching to each data entry a prescribed set of key words that a computer can search for matching items.

Mechanical devices to perform mathematical or logical operations have been around for centuries, but it was the invention of the electronic computer that revolutionized information processing. One aspect of mathematical logic is that any information whatsoever—including numbers, letters, and logical propositions—can be coded as a string of yes-or-no bits (for example, as dots and dashes, 1's and 0's, or on/off switches). Electronic computers are essentially very large arrays of on/off switches connected in ways that allow them to perform logical operations. New materials and techniques have made possible the extreme miniaturization and reliability of no-moving-parts switches, which enable very large numbers of connected switches to be fitted into a small space. Very small size also means very short connections, which in turn mean very brief travel time for signals; therefore, miniaturized electronic circuits can act very quickly. The very short times required for processing steps to occur, together with the very large number of connections that can be made, mean that computers can carry out extremely complicated or repetitive instructions millions of times more quickly than people can.

The activity of computers is controlled partly by how they are wired, partly by sets of coded instructions. In general-purpose computers, instructions for processing information are not wired in but are stored temporarily (like other information). This arrangement permits great flexibility in what computers can do. People give instructions to computers through previously programmed software or by means of original programs written in a programming language. Programming languages enable a programmer to compose instructions with something like English or algebra, or geometrical manipulation of diagrams. Those instructions are then translated by another program into machine language for the computer. Often, the program calls for other inputs in the form of data entered by keyboard, from an information-storage device, or from an automatic sensing device. The output of a computer may be symbolic (words, numbers) or graphic (charts, diagrams), or it may be the automatic control of some other machine (an alarm signal, an action of a robot) or a request to a human operator for more instructions.

An important role of computers is in modeling or simulating systems—for example, the economy or the weather, a grid of traffic lights, a strategic game, or chemical inter-

actions. In effect, the computer computes the logical consequences of a set of complicated instructions that represent how the system works. A computer program is written that specifies those instructions and is then run, beginning with data that describe an initial state of the system. The program also displays subsequent states of the system, which can be compared to how the systems actually behave to see how good our knowledge of the rules is and to help correct them. If we are sure we know all the rules well, we can use the consequence-deducing power of computers to aid us in the design of systems.

An important potential role for computer programs is to assist humans in problem solving and decision making. Computers already play a role in helping people think by running programs that amass, analyze, summarize, and display data. Pattern-searching programs help to extract meaning from large pools of data. An important area of research in computer science is the design of programs—based on the principles of artificial intelligence—that are intended to mimic human thought and possibly even improve on it. Most of human thought is not yet well understood, however. As is true for simulations of other complex systems such as the economy or the weather, comparison of the performance of programs with the phenomena they represent is a technique for learning more about how the system works.

In mechanical systems that are well understood, computers can provide control that is as good as, or more precise and rapid than, deliberate human control. Thus, the operation of automobile engines, the flight control of aircraft and spacecraft, and the aiming and firing of weapons can be computerized to take account of more information and to respond much more rapidly than a human operator could. Yet, there are also risks that the instructions or the information entered may contain errors, the computer may have malfunctions in its hardware or software, and even that perfectly reliable computers, programs, and information may still give faulty results if some relevant factors are not included in the programs or if any values of included factors fall outside of their expected range. Even if the whole system is technically flawless, though, a very complex high-speed system may create problems because its speed of response may exceed human ability to monitor or judge the output.

The complexity of control in today's world requires vast computer management of information. And as the amount of information increases, there is increasing need to keep track of, control, and interpret the information—which involves still more information, and so on through more layers of information. This flood of information requires invention of ways to store it in less space, to categorize it more usefully, to retrieve it more quickly, to transmit it at a higher rate, to sort it and search it more efficiently, and to minimize errors—that is, to check for them and to correct them when they are found. As for communication, information storage also involves issues of privacy and security. Computer-managed information systems require means for ensuring that information cannot be changed or lost accidentally and that it will be unintelligible if unauthorized access does occur.

HEALTH TECHNOLOGY

Health technology is concerned with reducing the exposure of humans to conditions that threaten health, as well as with increasing the body's resistance to such conditions and minimizing the deleterious effects that do occur.

Historically, the most important effect of technology on human health has been through prevention of disease, not through its treatment or cure. The recognition that disease-causing organisms are spread through insects, rodents, and human waste has led to great improvements in sanitation that have greatly affected the length and quality of human life. Sanitation measures include containment and disposal of garbage, construction of sewers and sewage-processing plants, purification of water and milk supplies, quarantine of infectious patients (or adequate prophylaxis), chemical reduction of insect and microorganism populations (insecticides and antiseptics), and suppression of the population of rats, flies, and mosquitoes that carry microorganisms. Just as important in the prevention of illness has been furnishing an adequate food supply containing the variety of foods required to supply all the body's needs for proteins, minerals, and trace substances.

Health technology can be used to enhance the human body's natural defenses against disease. Under conditions of reasonably good nutrition and sanitation, the human body recovers from most infectious diseases by itself without intervention of any kind, and recovery itself often brings immunity. But the suffering and danger of many serious diseases can be prevented artificially. By means of inoculation, the immune system of the human body can be provoked to develop its own defenses against specific disease without the suffering and risk of actually contracting the disease. Weakened or killed disease microorganisms injected into the blood may arouse the body's immune system to create antibodies that subsequently will incapacitate live microorganisms if they try to invade. Next to sanitation, inoculation has been the most effective means of preventing early death from disease, especially among infants and children.

Molecular biology is beginning to make it possible to design substances that evoke immune responses more precisely and safely than current vaccines. Genetic engineering is developing ways to induce organisms to produce these substances in quantities large enough for research and applications.

Many diseases are caused by bacteria or viruses. If the body's immune system cannot suppress a bacterial infection, an antibacterial drug may be effective—at least against the types of bacteria it was designed to combat. But the overuse of any given antibacterial drug can lead, by means of natural selection, to the spread of bacteria that are not affected by it. Much less is known about the treatment of viral infections, and there are very few antiviral drugs equivalent to those used to combat bacterial infections.

The detection, diagnosis, and monitoring of disease are improved by several different kinds of technology. Considerable progress in learning about the general condition of the human body came with the development of simple mechanical devices for measuring temperature and blood pressure and listening to the heart. A better look inside the body has been provided by imaging devices that use slender probes to supply visible light or (from outside the body) magnetic fields, infrared radiation, sound waves, x rays, or nuclear radiation. Using mathematical models of wave behavior, computers are able to process information from these probes to produce moving, three-dimensional images. Other technologies include chemical techniques for detecting disease-related components of body fluids and comparing levels of common components to normal ranges.

Techniques for mapping the location of genes on chromosomes make it possible to detect disease-related genes in children or in prospective parents; the latter can be informed and counseled about possible risks. With ever-growing technology for observing and measuring the body, the information load can become more than doctors can easily consider all at once. Computer programs that compare a patient's data to norms and to patterns typical of disease are increasingly aiding in diagnosis.

The modern treatment of many diseases also is improved by science-based technologies. Knowledge of chemistry, for example, has improved our understanding of how drugs and naturally occurring body chemicals work, how to synthesize them in large quantities, and how to supply the body with the proper amounts of them. Substances have been identified that are most damaging to certain kinds of cancer cells. Knowledge of the biological effects of finely controlled beams of light, ultrasound, x rays, and nuclear radiation (all at much greater intensities than are used for imaging) has led to technological alternatives to scalpels and cauterization. As the knowledge of the human immune system has grown and new materials have been developed, the transplantation of tissue or whole organs has become increasingly common. New materials that are durable and not rejected by the immune system now make it possible to replace some body parts and to implant devices for electrically pacing the heart, sensing internal conditions, or slowly dispensing drugs at optimal times.

Effective treatment of mental disturbance involves attention not only to the immediate psychological symptoms but also to their possible physiological causes and consequences, and to their possible roots in the individual's total experience. Psychological treatment may include prolonged or intensive personal interviews, group discussions among people with similar problems, or deliberately programmed punishment and reward to shape behavior. Medical treatment may include the use of drugs, or electric shock, or even surgery. The overall effectiveness of any of these treatments, even more than in the case of most other medical treatment, is uncertain; any one approach may work in some cases and not in others.

Improved medical technologies raise ethical and economic issues. The combined results of improved technology in public health, medicine, and agriculture have increased human longevity and population size. This growth in numbers, which is very unlikely to end before the middle of the next century, increases the challenge of providing all humans with adequate food, shelter, health care, and employment, and it places ever more strain on the environment. The high costs of some treatments forces society to make unwelcome choices about who should be selected to benefit and who should pay. Moreover, the developing technology of diagnosing, monitoring, and treating diseases and malfunctions increases society's ability to keep people living when they otherwise would have been unable to sustain their lives themselves. This raises questions as to who should decide whether and for how long extraordinary care should be provided and to whom. There is continuing debate over abortion, intensive care for infants with severe disabilities, maintaining the life functions of people whose brains have died, the sale of organs, altering human genes, and many other social and cultural issues that arise from biomedical technology.

An increasingly important adjunct to preventive and corrective health care is the use of statistics to keep track of the distribution of disease, malnutrition, and death among various geographic, social, and economic groups. They help determine where public health problems are and how fast they may be spreading. Such information can be interpreted, sometimes with the help of mathematical modeling, to project the effects of preventive and corrective measures and thus to plan more effectively.

Stuart Davis, *Rapt at Rappaport's* (1952).

CHAPTER 9

THE MATHEMATICAL WORLD

Mathematics is essentially a process of thinking that involves building and applying abstract, logically connected networks of ideas. These ideas often arise from the need to solve problems in science, technology, and everyday life—problems ranging from how to model certain aspects of a complex scientific problem to how to balance a checkbook.

This chapter presents recommendations about basic mathematical ideas, especially those with practical application, that together play a key role in almost all human endeavors. In Chapter 2, mathematics is characterized as a modeling process in which abstractions are made and manipulated and the implications are checked out against the original situation. Here, the focus is on seven examples of the kinds of mathematical patterns that are available for such modeling: the nature and use of numbers, symbolic relationships, shapes, uncertainty, summarizing data, sampling data, and reasoning.

RECOMMENDATIONS

NUMBERS

There are several kinds of numbers that in combination with a logic for interrelating them form interesting abstract systems and can be useful in a variety of very different ways. The age-old concept of number probably originated in the need to count how many things there were in a collection of things. Thus, fingers, pebbles in containers, marks on clay tablets, notches on sticks, and knots on cords were all early ways of keeping track of and representing counted quantities. More recently, during the past 2,000 years or so, various systems of writing have been used to represent numbers. The Arabic number system, as commonly used today, is based on ten symbols (0, 1, 2, . . . 9) and rules for combining them in which position is crucial (for example, in 203, the 3 stands for three, the 2 stands for two hundreds, and the zero stands for no additional tens). In the binary system—the mathematical language of computers—just two symbols, 0 and 1, can be combined in a string to represent any number. The Roman number system, which is still used for some purposes (but rarely for calculation), is made up of a few letters of the alphabet and rules for combining them (for example, IV for four, X for ten, and XIV for fourteen, but no symbol for zero).

There are different kinds of numbers. The numbers that come from counting things are whole numbers, which are the numbers we mostly use in everyday life. A whole number by itself is an abstraction for how many things there are in a set but not for the things themselves. "Three" can refer to apples, boulders, people, volts, miles per hour, or anything else. But in most practical situations, we want to know what the objects are, as well as how many there are. Thus, the answer to most calculations is a magnitude—a number connected to a label. If some people traveled 165 miles in 3 hours, their average speed was 55 miles per hour, not 55. In this instance, 165, 3, and 55 are numbers; 165 miles, 3 hours, and 55 miles per hour are magnitudes. The labels are important in keeping track of the meanings of the numbers.

Fractions are numbers we use to stand for a part of something or to compare two quantities. One common kind of comparison occurs when some magnitude such as length or weight is measured—that is, is compared to a standard unit such as a meter or a pound. Two kinds of symbols are widely used to stand for fractions, but they are numerically equivalent. For example, the ordinary fraction ¾ and the decimal fraction 0.75 both represent the same number. Used to represent measured magnitudes, however, the two expressions may have somewhat different implications: ¾ could be used to simply mean closer to ¾ than to ²/₄ or ⁴/₄, whereas 0.75 may imply being closer to 0.75 than to 0.74 or 0.76—a much more precise specification. Whole numbers and fractions can be used together: 1¼, 1.25, ¹²⁵/₁₀₀, and ⁵/₄, for instance, all mean the same thing numerically.

More flexibility in mathematics is provided by the use of negative numbers, which can be thought of in terms of a number line. A number line lays consecutive numbers at equal intervals along a straight line centered on zero. The numbers on one side of zero are called positive, and those on the other side, negative. If the numbers to the right of zero are positive, the numbers to the left of zero are negative; if distance above sea level is positive, distance below sea level is negative; if income is positive, debt is negative. If 2:15 is the scheduled time of lift-off, 2:10 is "minus 5 minutes." The complete range of numbers—positive, zero, and negative—allows any number to be subtracted from any other and still give an answer.

Computation is the manipulation of numbers and other symbols to arrive at some new mathematical statement. These other symbols may be letters used to stand for numbers. For example, in trying to solve a particular problem, we might let X stand for any number that would meet the conditions of the problem. There are also symbols to signify what operations to perform on the number symbols. The most common ones are $+$, $-$, $\times$, and $\div$ (there are also others). The operations $+$ and $-$ are inverses of each other, just as $\times$ and $\div$ are: One operation undoes what the other does. The expression a/b can mean "the quantity a compared to the quantity b," or "the number you get if you divide a by b," or "a parts of size $1/b$." The parentheses in $a(b + c)$ tell us to multiply a by the sum of b and c. Mathematicians study systems of numbers to discover their properties and relationships and to devise rules for manipulating mathe-

matical symbols in ways that give valid results.

Numbers have many different uses, some of which are not quantitative or strictly logical. In counting, for example, zero has a special meaning of nothing. Yet, on the common temperature scale, zero is only an arbitrary position and does not mean an absence of temperature (or of anything else). Numbers can be used to put things in an order and to indicate only which is higher or lower than others—not to specify by how much (for example, the order of winners in a race, street addresses, or scores on psychological tests for which numerical differences have no uniform meaning). And numbers are commonly used simply to identify things without any meaningful order, as in telephone numbers and as used on athletic shirts and license plates.

Aside from their application to the world of everyday experience, numbers themselves are interesting. Since earliest history, people have asked such questions as, Is there a largest number? A smallest number? Can every possible number be obtained by dividing some whole number by another? And some numbers, such as the ratio of a circle's circumference to its diameter (pi), catch the fancy of many people, not just mathematicians.

SYMBOLIC RELATIONSHIPS

Numbers and relationships among them can be represented in symbolic statements, which provide a way to model, investigate, and display real-world relationships. Seldom are we interested in only one quantity or category; rather, we are usually interested in the relationship between them—the relationship between age and height, temperature and time of day, political party and annual income, sex and occupation. Such relationships can be expressed by using pictures (typically charts and graphs), tables, algebraic equations, or words. Graphs are especially useful in examining the relationships between quantities.

Algebra is a field of mathematics that explores the relationships among different quantities by representing them as symbols and manipulating statements that relate the symbols. Sometimes a symbolic statement implies that only one value or set of values will make the statement true. For example, the statement $2A + 4 = 10$ is true if (and only if) $A = 3$. More generally, however, an algebraic statement allows a quantity to take on any of a range of values and implies for each what the corresponding value of another quantity is. For example, the statement $A = s^2$ specifies a value for the variable A that corresponds to any choice of a value for the variable s.

There are many possible kinds of relationships between one variable and another. A basic set of simple examples includes (1) directly proportional (one quantity always keeps the same proportion to another), (2) inversely proportional (as one quantity increases, the other decreases proportionally), (3) accelerated (as one quantity increases uniformly, the other increases faster and faster), (4) converging (as one quantity increases without limit, the other approaches closer and closer to some limiting value), (5) cyclical (as one quantity increases, the other increases and decreases in repeating cycles), and (6) stepped (as one quantity changes smoothly, the other changes in jumps).

Symbolic statements can be manipulated by rules of mathematical logic to produce other statements of the same relationship, which may show some interesting aspect more clearly. For example, we could state symbolically the relationship between the width of a page, P, the length of a line of type, L, and the width of each vertical margin, m: $P - L + 2m$. This equation is a useful model for determining page makeup. It can be rearranged logically to give other true statements of the same basic relationship: for example, the equations $L = P - 2m$ or $m = (P - L)/2$, which may be more convenient for computing actual values for L or m.

In some cases, we may want to find values that will satisfy two or more different relationships at the same time. For example, we could add to the page-makeup model another condition: that the length of the line of type must be ⅔ of the page width: $L = ⅔P$. Combining this equation with $m = (P - L)/2$, we arrive logically at the result that $m = ⅙ P$. This new equation, derived from the other two together, specifies the only values for m that will fit both relationships. In this simple example, the specification for the margin width could be

worked out readily without using the symbolic relationships. In other situations, however, the symbolic representation and manipulation are necessary to arrive at a solution—or to see whether a solution is even possible.

Often, the quantity that interests us most is how fast something is changing rather than the change itself. In some cases, the rate of change of one quantity depends on some other quantity (for example, change in the velocity of a moving object is proportional to the force applied to it). In some other cases, the rate of change is proportional to the quantity itself (for example, the number of new mice born into a population of mice depends on the number and gender of mice already there).

SHAPES

Spatial patterns can be represented by a fairly small collection of fundamental geometrical shapes and relationships that have corresponding symbolic representation. To make sense of the world, the human mind relies heavily on its perception of shapes and patterns. The artifacts around us (such as buildings, vehicles, toys, and pyramids) and the familiar forms we see in nature (such as animals, leaves, stones, flowers, and the moon and sun) can often be characterized in terms of geometric form. Some of the ideas and terms of geometry have become part of everyday language. Although real objects never perfectly match a geometric figure, they more or less approximate them, so that what is known about geometric figures and relationships can be applied to objects. For many purposes, it is sufficient to be familiar with points, lines, planes; triangles, rectangles, squares, circles, and ellipses; rectangular solids and spheres; relationships of similarity and congruence; relationships of convex, concave, intersecting, and tangent; angles between lines or planes; parallel and perpendicular relationships between lines and planes; forms of symmetry such as displacement, reflection, and rotation; and the Pythagorean theorem.

Both shape and scale can have important consequences for the performance of systems. For example, triangular connections maximize rigidity, smooth surfaces minimize turbulence, and a spherical container minimizes surface area for any given mass or volume. Changing the size of objects while keeping the same shape can have profound effects owing to the geometry of scaling: Area varies as the square of linear dimensions, and volume varies as the cube. On the other hand, some particularly interesting kinds of patterns known as fractals look very similar to one another when observed at any scale whatever—and some natural phenomena (such as the shapes of clouds, mountains, and coastlines) seem to be like that.

Geometrical relationships can also be expressed in symbols and numbers, and vice versa. Coordinate systems are a familiar means of relating numbers to geometry. For the simplest example, any number can be represented as a unique point on a line—if we first specify points to represent zero and one. On any flat surface, locations can be specified uniquely by a pair of numbers or coordinates. For example, the distance from the left side of a map and the distance from the bottom, or the distance and direction from the map's center.

Coordinate systems are essential to making accurate maps, but there are some subtleties. For example, the approximately spherical surface of the earth cannot be represented on a flat map without distortion. Over a few dozen miles, the problem is barely noticeable; but on the scale of hundreds or thousands of miles, distortion necessarily appears. A variety of approximate representations can be made, and each involves a somewhat different kind of distortion of shape, area, or distance. One common type of map exaggerates the apparent areas of regions close to the poles (for example, Greenland and Alaska), whereas other useful types misrepresent what the shortest distance between two places is, or even what is adjacent to what.

Mathematical treatment of shape also includes graphical depiction of numerical and symbolic relationships. Quantities are visualized as lengths or areas (as in bar and pie charts) or as distances from reference axes (as in line graphs or scatter plots). Graphical display makes it possible to readily identify patterns that might not otherwise be obvious: for example, relative sizes (as proportions or differences), rates of change (as slopes), abrupt discontinuities (as gaps or jumps), clustering (as distances between plotted points), and trends (as changing slopes or projections). The mathematics of geometric relations also aids in analyzing the design of complex structures (such as protein molecules or airplane wings) and logical networks (such as connections of brain cells or long-distance telephone systems).

UNCERTAINTY

Our knowledge of how the world works is limited by at least five kinds of uncertainty: (1) inadequate knowledge of all the factors that may influence something, (2) inadequate number of observations of those factors, (3) lack of precision in the observations, (4) lack of appropriate models to combine all the information meaningfully, and (5) inadequate ability to compute from the models. It is possible to predict some events with great accuracy (eclipses), others with fair accuracy (elections), and some with very little certainty (earthquakes). Although absolute certainty is often impossible to attain, we can often estimate the likelihood—whether large or small—that some things will happen and what the likely margin of error of the estimate will be.

It is often useful to express likelihood as a numerical probability. We usually use a probability scale of 0 to 1, where 0 indicates our belief that some particular event is certain not to occur, 1 indicates our belief that it is certain to occur, and anything in between indicates uncertainty. For example, a probability of .9 indicates a belief that there are 9 chances in 10 of an event occurring as predicted; a probability of .001 indicates a belief that there is only 1 chance in 1,000 of its occurring. Equivalently, probabilities can also be expressed as percentages, ranging from 0 percent (no chance) to 100 percent (certainty). Uncertainties can also be expressed as odds: A probability of .8 for an event can be expressed as odds of 8 to 2 (or 4 to 1) in favor of its occurring.

One way to estimate the probability of an event is to consider past events. If the current situation is similar to past situations, then we may expect somewhat similar results. For example, if it rained on 10 percent of summer days last year, we could expect that it will rain on approximately 10 percent of summer days this year. Thus, a reasonable estimate for the probability of rain on any given summer day is .1—one chance in ten. Additional information can change our estimate of the probability. For example, rain may have fallen on 40 percent of the cloudy days last summer; thus, if our given day is cloudy, we would raise the estimate from .1 to .4 for the probability of rain. The more ways in which the situation we are interested in is like those for which we have data, the better our estimate is likely to be.

Another approach to estimating probabilities is to consider the possible alternative outcomes to a particular event. For example, if there are 38 equally wide slots on a roulette wheel, we may expect the ball to fall in each slot about $\frac{1}{38}$ of the time. Estimates of such a theoretical probability rest on the assumption that all of the possible outcomes are accounted for and all are equally likely to happen. But if that is not true—for example, if the slots are not of equal size or if sometimes the ball flies out of the wheel—the calculated probability will be wrong.

Probabilities are most useful in predicting proportions of results in large numbers of events. A flipped coin has a 50 percent chance of coming up heads, although a person will usually not get precisely 50 percent heads in an even number of flips. The more times one flips it, the less likely one is to get a count of precisely 50 percent but the closer the proportion of heads is likely to be to the theoretical 50 percent. Similarly, insurance companies can usually come within a percentage point or two of predicting the proportion of people aged 20 who will die in a given year but are likely to be off by thousands of total deaths—and they have no ability whatsoever to predict whether any particular 20-year-old will die. In other contexts, too, it is important to distinguish between the proportion and the actual count. When there is a very large number of similar events, even an outcome with a very small probability of occurring can occur fairly often. For example, a medical test with a probability of 99 percent of being correct may seem highly accurate—but if that test were performed on a million people, approximately 10,000 individuals would receive false results.

SUMMARIZING DATA

Information is all around us—often in such great quantities that we are unable to make sense of it. A set of data can be represented by a few summary characteristics that may reveal or conceal important aspects of it. Statistics is a form of mathematics that develops useful ways for organizing and analyzing large amounts of data. To get an idea of what a set of data is like, for example, we can plot each case on a number line, and then inspect the plot to see where cases are piled up, where some are separate from the others, where the highest and lowest are, and so on. Alternatively, the data set can be characterized in a summary fashion by describing where its middle is and how much variation there is around that middle.

The most familiar statistic for summarizing a data distribution is the mean, or common average; but care must be taken in using or interpreting it. When data are discrete (such as number of children per family), the mean may not even be a possible value (for example, 2.2 children). When data are highly skewed toward one extreme, the mean may not even be close to a typical value. For example, a small fraction of people who have very large personal incomes can raise the mean considerably higher than the bulk of people piled at the lower end can lower it. The median, which divides the lower half of the data from the upper half, is more meaningful for many purposes. When there are only a few discrete values of a quantity, the most informative kind of average may be the mode, which is the most common single value—for example, the most common number of cars per U.S. family is 1.

More generally, averages by themselves neglect variation in the data and may imply more uniformity than exists. For example, the average temperature on the planet Mercury of about 15° F does not sound too bad—until one considers that it swings from 300° F above to almost 300° F below zero. The neglect of variation can be particularly misleading when averages are compared. For example, the fact that the average height of men is distinctly greater than that of women could be reported as "men are taller than women," whereas many women are taller than many men. To interpret averages, therefore, it is important to have information about the variation within groups, such as the total range of data or the range covered by the middle 50 percent. A plot of all the data along a number line makes it possible to see how the data are spread out.

We are often presented with summary data that purport to demonstrate a relationship between two variables but lack essential information. For example, the claim that "more than 50 percent of married couples who have different religions eventually get divorced" would not tell us anything about the relationship between religion and divorce unless we also knew the percentage of couples with the *same* religion who get divorced. Only the comparison of the two percentages could tell us whether there may be a real relationship. Even then, caution is necessary because of possible bias in how the samples were selected and because differences in percentage could occur just by chance in selecting the sample. Proper reports of such information should include a description of possible sources of bias and an estimate of the statistical uncertainty in the comparison.

Two quantities are positively correlated if having more of one is associated with having more of the other. (A negative correlation means that having more of one is associated with having *less* of the other.) But even a strong correlation between two quantities does not mean that one is necessarily a cause of the other. Either one could possibly cause the other, or both could be the common result of some third factor. For example, life expectancy in a community is positively correlated with the average number of telephones per household. One could look for an explanation for how having more telephones improves one's health or why healthier people buy more telephones. More likely, however, both health and number of telephones are the consequence of the community's general level of wealth, which affects the overall quality of nutrition and medical care, as well as the people's inclination to buy telephones.

SAMPLING

Most of what we learn about the world is obtained from information based on samples of what we are studying—samples of,

say, rock formations, light from stars, television viewers, cancer patients, whales, or numbers. Samples are used because it may be impossible, impractical, or too costly to examine all of something, and because a sample often is sufficient for most purposes. In drawing conclusions about all of something from samples of it, two major concerns must be taken into account. First, we must be alert to possible bias created by how the sample was selected. Common sources of bias in drawing samples include convenience (for example, interviewing only one's friends or picking up only surface rocks), self-selection (for example, studying only people who volunteer or who return questionnaires), failure to include those who have dropped out along the way (for example, testing only students who stay in school or only patients who stick with a course of therapy), and deciding to use only the data that support our preconceptions.

A second major concern that determines the usefulness of a sample is its size. If sampling is done without bias in the method, then the larger the sample is, the more likely it is to represent the whole accurately. This is because the larger a sample is, the smaller the effects of purely random variations are likely to be on its summary characteristics. The chance of drawing a wrong conclusion shrinks as the sample size increases. For example, for samples chosen at random, finding that 600 out of a sample of 1,000 have a certain feature is much stronger evidence that a majority of the population from which it was drawn have that feature than finding that 6 out of a sample of 10 (or even 9 out of the 10) have it. On the other hand, the actual size of the total population from which a sample is drawn has little effect on the accuracy of sample results. A random sample of 1,000 would have about the same margin of error whether it were drawn from a population of 10,000 or from a similar population of 100 million.

REASONING

Some aspects of reasoning have clear logical rules, others have only guidelines, and still others have almost unlimited room for creativity (and, of course, error). A convincing argument requires both true statements and valid connections among them. Yet formal logic concerns the validity of the connections among statements, not whether the statements are actually true. It is logically correct to argue that if all birds can fly and penguins are birds, then penguins can fly. But the conclusion is not true unless the premises are true: Do all birds really fly, and are penguins really birds? Examination of the truth of premises is as important to good reasoning as the logic that operates on them is. In this case, because the logic is correct but the conclusion is false (penguins cannot fly), one or both of the premises must be false (not all birds can fly, and/or penguins are not birds).

Very complex logical arguments can be built from a small number of logical steps, which hang on precise use of the basic terms "if," "and," "or," and "not." For example, medical diagnosis involves branching chains of logic such as "If the patient has disease X or disease Y and also has laboratory result B, but does not have a history of C, then he or she should get treatment D." Such logical problem solving may require expert knowledge of many relationships, access to much data to feed into the relationships, and skill in deducing branching chains of logical operations. Because computers can store and retrieve large numbers of relationships and data and can rapidly perform long series of logical steps, they are being used increasingly to help experts solve complex problems that would otherwise be very difficult or impossible to solve. Not all logical problems, however, can be solved by computers.

Logical connections can easily be distorted. For example, the proposition that all birds can fly does not imply logically that all creatures that can fly are birds. As obvious as this simple example may seem, distortion often occurs, particularly in emotionally charged situations. For example: All guilty prisoners refuse to testify against themselves; prisoner Smith refuses to testify against himself; therefore, prisoner Smith is guilty.

Distortions in logic often result from not distinguishing between necessary conditions and sufficient conditions. A condition that is necessary for a consequence is always required but may not be enough in itself—being a U.S. citizen is necessary to be elected president, for example, but not sufficient. A condition that is sufficient for a consequence is enough by itself, but there may be other ways to arrive at the same consequence—winning the state lottery is sufficient for becoming rich, but there are other ways. A condition, however, may be both necessary and sufficient; for example, receiving a majority of the electoral vote is both necessary for becoming president and sufficient for doing so, because it is the only way.

Logic has limited usefulness in finding solutions to many problems. Outside of abstract models, we often cannot establish with confidence either the truth of the premises or the logical connections between them. Precise logic requires that we can make declarations such as "If X is true, then Y is true also" (a barking dog does not bite), and "X is true" (Spot barks). Typically, however, all we know is that "if X is true, then Y is often true also" (a barking dog usually does not bite) and "X seems to be approximately true a lot of the time" (Spot usually barks). Commonly, therefore, strict logic has to be replaced by probabilities or other kinds of reasoning that lead to much less certain results—for example, to the claim that on average, rain will fall before evening on 70 percent of days that have morning weather conditions similar to today's.

If we apply logical deduction to a general rule (all feathered creatures fly), we can produce a conclusion about a particular instance or class of instances (penguins fly). But where do the general rules come from? Often they are generalizations made from observations—finding a number of similar instances and guessing that what is true of them is true of all their class ("every feathered creature I have seen can fly, so perhaps all can"). Or a general rule may spring from the imagination, by no traceable means, with the hope that some observable aspects of phenomena can be shown to follow logically from it (example: "What if it were true that the sun is the center of motion for all the planets, including the earth? Could such a system produce the apparent motions in the sky?").

Once a general rule has been hypothesized, by whatever means, logic serves in checking its validity. If a contrary instance is found (a feathered creature that cannot fly), the hypothesis is not true. On the other hand, the only way to prove logically that a general hypothesis about a class is true is to examine all possible instances (all birds), which is difficult in practice and sometimes impossible even in principle. So it is usually much easier to prove general hypotheses to be logically false than to prove them to be true. Computers now sometimes make it possible to demonstrate the truth of questionable mathematical generalizations convincingly, even if not to prove them, by testing enormous numbers of particular cases.

Science can use deductive logic if general principles about phenomena have been hypothesized, but such logic cannot lead to those general principles. Scientific principles are usually arrived at by generalizing from a limited number of experiences—for instance, if all observed feathered creatures hatch from eggs, then perhaps all feathered creatures do. This is a very important kind of reasoning even if the number of observations is small (for example, being burned once by fire may be enough to make someone wary of fire for life). However, our natural tendency to generalize can also lead us astray. Getting sick the day after breaking a mirror may be enough to make someone afraid of broken mirrors for life. On a more sophisticated level, finding that several patients having the same symptoms recover after using a new drug may lead a doctor to generalize that all similar patients will recover by using it, even though recovery might have occurred just by chance.

The human tendency to generalize has some subtle aspects. Once formed, generalities tend to influence people's perception and interpretation of events. Having made the generalization that the drug will help all patients having certain symptoms, for example, the doctor may be likely to interpret a patient's condition as having improved after taking the drug, even if that is doubtful. To prevent such biases in research, scientists commonly use a "blind" procedure in which the person observing or interpreting results is not the same person who controls the conditions (for instance, the doctor who judges the patient's condition does not know what specific treatment that patient received).

Much of reasoning, and perhaps most of creative thought, involves not only logic but analogies. When one situation seems to resemble another in some way, we may believe that it resembles it in other ways too. For example, light spreads away from a source much as water waves spread from a disturbance, so perhaps light acts like water waves in other ways, such as producing interference patterns where waves cross (it does). Or, the sun is like a fire in that it produces heat and light, so perhaps it too involves burning fuel (in fact, it does not). The important point is that reasoning by analogy can suggest conclusions, but it can never prove them to be true.

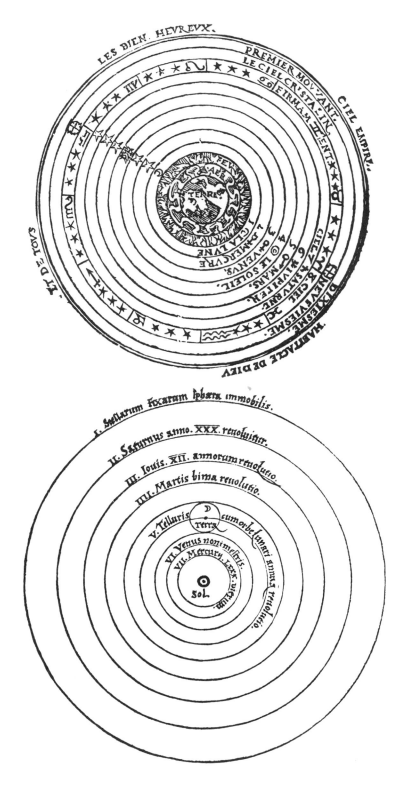

Earth-centered and sun-centered models of the solar system
as depicted in the sixteenth century.

CHAPTER 10
HISTORICAL PERSPECTIVES

There are two principal reasons for including some knowledge of history among the recommendations. One reason is that generalizations about how the scientific enterprise operates would be empty without concrete examples. Consider, for example, the proposition that new ideas are limited by the context in which they are conceived; are often rejected by the scientific establishment; sometimes spring from unexpected findings; and usually grow slowly, through contributions from many different investigators. Without historical examples, these generalizations would be no more than slogans, however well they might be remembered. For this purpose, any number of episodes might have been selected.

A second reason is that some episodes in the history of the scientific endeavor are of surpassing significance to our cultural heritage. Such episodes certainly include Galileo's role in changing our perception of our place in the universe; Newton's demonstration that the same laws apply to motion in the heavens and on earth; Darwin's long observations of the variety and relatedness of life forms that led to his postulating a mechanism for how they came about; Lyell's careful documentation of the unbelievable age of the earth; and Pasteur's identification of infectious disease with tiny organisms that could be seen only with a microscope. These stories stand among the milestones of the development of all thought in Western civilization.

The recommendations in this chapter focus on Western cultural development, rather than on the many ideas in science and mathematics and the technological inventions that originated in early Egyptian, Greek, Chinese, and Arabic cultures. The sciences accounted for in this report are largely part of a tradition of thought that happened to develop in Europe during the last 500 years—a tradition to which people from all cultures contribute today.

The emphasis here is on ten accounts of significant discoveries and changes that exemplify the evolution and impact of scientific knowledge: the planetary earth, universal gravitation, relativity, geologic time, plate tectonics, the conservation of matter, radioactivity and nuclear fission, the evolution of species, the nature of disease, and the Industrial Revolution. Although other choices may be equally valid, these clearly fit our dual criteria of exemplifying historical themes and having cultural salience.

RECOMMENDATIONS

DISPLACING THE EARTH FROM THE CENTER OF THE UNIVERSE

To observers on the earth, it appears that the earth stands still and everything else moves around it. Thus, in trying to imagine how the universe works, it made good sense to people in ancient times to start with those apparent truths. The ancient Greek thinkers, particularly Aristotle, set a pattern that was to last for about 2,000 years: a large, stationary earth at the center of the universe, and—positioned around the earth— the sun, the moon, and tiny stars arrayed in a perfect sphere, with all these bodies orbiting along perfect circles at constant speeds. Shortly after the beginning of the Christian era, that basic concept was transformed into a powerful mathematical model by an Egyptian astronomer, Ptolemy. His model of perfect circular motions served well for predicting the positions of the sun, moon, and stars. It even accounted for some motions in the heavens that appeared distinctly irregular. A few "wandering stars"—the planets—appeared not to circle perfectly around the earth but rather to change speed and sometimes even go into reverse, following odd loop-the-loop paths. This behavior was accounted for in Ptolemy's model by adding more circles, which spun on the main circles.

Over the following centuries, as astronomical data accumulated and became more accurate, this model was refined and complicated by many astronomers, including Arabs and Europeans. As clever as the refinements of perfect-circles models were, they did not involve any physical explanations of why heavenly bodies should so move. The principles of motion in the heavens were considered to be quite different from those of motion on earth.

Shortly after the discovery of the Americas, a Polish astronomer named Nicolaus Copernicus, a contemporary of Martin Luther and Leonardo da Vinci, proposed a different model of the universe. Discarding the premise of a stationary earth, he showed that if the earth and planets all circled around the sun, the apparent erratic motion of the planets could be accounted for just as well, and in a more intellectually pleasing way. But Copernicus' model still used perfect circular motions and was nearly as complicated as the old earth-centered model. Moreover, his model violated the prevailing common-sense notions about the world, in that it required the apparently immobile earth to spin completely around on its axis once a day, the universe to be far larger than had been imagined, and—worst of all—the earth to become commonplace by losing its position at the center of the universe. Further, an orbiting and spinning earth was thought to be inconsistent with some biblical passages. Most scholars perceived too little advantage in a sun-centered model—and too high a cost in giving up the many other ideas associated with the traditional earth-centered model.

As astronomical measurements continued to become more precise, it became clear that neither the sun-centered nor the earth-centered system quite worked as long as all bodies had to have uniform circular motion. A German astronomer, Johannes Kepler, who lived at the same time as Galileo, developed a mathematical model of planetary motion that discarded both venerable premises—a stationary earth and circular motion. He postulated three laws, the most revolutionary of which was that planets naturally move in elliptical orbits at predictable but varying speeds. Although this law turned out to be correct, the calculations for ellipses were difficult with the mathematics known at the time, and Kepler offered no explanation for why the planets would move that way.

The many contributions of Italian scientist Galileo, who lived at the same time as Shakespeare and Rubens, were of great significance in the development of physics and astronomy. As an astronomer, he built and used the newly invented telescope to study the sun, moon, planets, and stars, and he made a host of discoveries that supported Copernicus' basic idea of planetary movement. Perhaps the most telling of these was his discovery of four moons that orbited around the planet Jupiter, demonstrating that the earth was not the only center of heavenly motion. With the telescope, he also discovered the inexplicable phenomena of craters and mountains on the moon, spots on the sun, moonlike phases of Venus, and vast numbers of stars not visible to the unaided eye.

Galileo's other great contribution to the cosmological revolution was in taking it to the public. He presented the new view in a form and language (Italian) that made it accessible to all educated people in his time. He also rebutted many popular arguments against an orbiting and spinning earth and showed inconsistencies in the Aristotelian account of motion. Criticism from clergy who still believed in Ptolemy's model—and Galileo's subsequent trial by the Inquisition for his allegedly heretical beliefs—only heightened the attention paid to the issues and thereby accelerated the process of changing generally accepted ideas on what constituted common sense. It also revealed some of the inevitable tensions that are bound to occur whenever scientists come up with radically new ideas.

UNITING THE HEAVENS AND EARTH

But it remained for Isaac Newton, an English scientist, to bring all of those strands together, and go far beyond them, to create the idea of the new universe. In his *Mathematical Principles of Natural Philosophy*, published near the end of the seventeenth century and destined to become one of the most influential books ever written, Newton presented a seamless mathematical view of the world that brought together knowledge of the motion of objects on earth and of the distant motions of heavenly bodies.

The Newtonian world was a surprisingly simple one: Using a few key concepts (mass, momentum, acceleration, and force), three laws of motion (inertia, the dependence of acceleration on force and mass, and action and reaction), and the mathematical law of how the force of gravity between all masses depends on distance, Newton was able to give rigorous explanations for motion on the earth and in the heavens. With a single set of ideas, he was able to account for the observed orbits of planets and moons, the motion of comets, the irregular motion of the moon, the motion of falling objects at the earth's surface, weight, ocean tides, and the earth's slight equatorial bulge. Newton made the earth part of an understandable universe, a universe elegant in its simplicity and majestic in its architecture—a universe that ran automatically by itself according to the action of forces between its parts.

Newton's system prevailed as a scientific and philosophical view of the world for 200 years. Its early acceptance was dramatically ensured by the verification of Edmund Halley's prediction, made many years earlier, that a certain comet would reappear on a particular date calculated from Newton's principles. Belief in Newton's system was continually reinforced by its usefulness in science and in practical endeavors, right up to (and including) the exploration of space in the twentieth century. Albert Einstein's theories of relativity—revolutionary in their own right—did not overthrow the world of Newton, but modified some of its most fundamental concepts.

The science of Newton was so successful that its influence spread far beyond physics and astronomy. Physical principles and Newton's mathematical way of deriving consequences from them together became the model for all other sciences. The belief grew that eventually all of nature could be explained in terms of physics and mathematics and that nature therefore could run by itself, without the help or attention of gods—although Newton himself saw his physics as demonstrating the hand of God acting on the universe. Social thinkers considered whether governments could be designed like a Newtonian solar system, with a balance of forces and actions that would ensure regular operation and long-term stability.

Philosophers in and outside of science were troubled by the implication that if everything from stars to atoms runs according to precise mechanical laws, the human notion of free will might be only an illusion. Could all human history, from thoughts to social upheavals, be only the playing out of a completely determined sequence of events? Social thinkers raised questions about free will and the organization of social systems that were widely debated in the eighteenth and nineteenth centuries. In the twentieth century, the appearance of basic unpredictability in the behavior of atoms relieved some of those concerns—but also raised new philosophical questions.

UNITING MATTER AND ENERGY, TIME AND SPACE

As elaborate and successful as it was, however, the Newtonian world view finally had to undergo some fundamental revisions around the beginning of the twentieth century. Still only in his twenties, German-born Albert Einstein published theoretical ideas that made revolutionary contributions to the understanding of nature. One of these was the special theory of relativity, in which Einstein considered space and time to be closely linked dimensions rather than, as Newton had thought, to be completely different dimensions.

Relativity theory had several surprising implications. One is that the speed of light is measured to be the same by all observers, no matter how they or the source of light happen to be moving. This is not true for the motion of other things, for their measured speed always depends on the motion of the observer. Moreover, the speed of light in empty space is the greatest speed possible—nothing can be accelerated up to that speed or observed moving faster.

The special theory of relativity is best known for asserting the equivalence of mass and energy—that is, any form of energy has mass, and matter itself is a form of energy. This is expressed in the famous equation $E = mc^2$, in which E stands for energy, m for mass, and c for the speed of light. Since c is approximately 186,000 miles per second, the transformation of even a tiny amount of mass releases an enormous amount of energy. That is what happens in the nuclear fission reactions that produce heat energy in nuclear reactors, and also in the nuclear fusion reactions that produce the energy given off by the sun.

About a decade later, Einstein published what is regarded as his crowning achievement and one of the most profound accomplishments of the human mind in all of history: the theory of general relativity. The theory has to do with the relationship between gravity and time and space, in which Newton's gravitational force is interpreted as a distortion in the geometry of space and time. Relativity theory has been tested over and over again by checking predictions based on it, and it has never failed. Nor has a more powerful theory of the architecture of the universe replaced it. But many physicists are looking for ways to come up with a more complete theory still, one that will link general relativity to the quantum theory of atomic behavior.

EXTENDING TIME

The age of the earth was not at issue for most of human history. Until the nineteenth century, nearly everyone in Western cultures believed that the earth was only a few thousand years old, and that the face of the earth was fixed—the mountains, valleys, oceans, and rivers were as they always had been since their instantaneous creation. From time to time, individuals speculated on the possibility that the earth's surface had been shaped by the kind of slow change processes they could observe occurring; in that case, the earth might have to be older than most people believed. If valleys were formed from erosion by rivers, and if layered rock originated in layers of sediment from erosion, one could estimate that millions of years would have been required to produce today's landscape. But the argument made only very gradual headway until English geologist Charles Lyell published the first edition of his masterpiece, *Principles of Geology,* early in the nineteenth century. The success of Lyell's book stemmed from its wealth of observations of the patterns of rock layers in mountains and the locations of various kinds of fossils, and from the close reasoning he used in drawing inferences from those data.

Principles of Geology went through many editions and was studied by several generations of geology students, who came to accept Lyell's philosophy and to adopt his methods of investigation and reasoning. Moreover, Lyell's book also influenced Charles Darwin, who read it while on his worldwide voyages studying the diversity of species. As Darwin developed his concept of biological evolution, he adopted Lyell's premises about the age of the earth and Lyell's style of buttressing his argument with massive evidence.

As often happens in science, Lyell's revolutionary new view that so opened up thought about the world also came to restrict his own thinking. Lyell took the idea of very slow change to imply that the earth never changed in sudden ways—and in fact really never changed much in its general features at all, perpetually cycling through similar sequences of small-scale changes. However, new evidence continued to accumulate; by the middle of the twentieth century, geologists believed that such minor cycles were only part of a complex process that also included abrupt or even cataclysmic changes and long-term evolution into new states.

SETTING THE EARTH'S SURFACE IN MOTION

As soon as fairly accurate world maps began to appear, some people noticed that the continents of Africa and South America looked as though they might fit together, like a giant jigsaw puzzle. Could they once have been part of a single giant landmass that broke into pieces and then drifted apart? The idea was repeatedly suggested, but was rejected for lack of evidence. Such a notion seemed fanciful in view of the size, mass, and rigidity of the continents and ocean basins and their apparent immobility.

Early in the twentieth century, however, the idea was again introduced, by German scientist Alfred Wegener, with new evidence: The outlines of the underwater edges of continents fit together even better than the above-water outlines; the plants, animals, and fossils on the edge of one continent were like those on the facing edge of the matching continent; and—most important—measurements showed that Greenland and Europe were slowly moving farther apart. Yet the idea had little acceptance (and strong opposition) until—with the development of new techniques and instruments—still more evidence accumulated. Further matches of continental shelves and ocean features were found by exploration of the composition and shape of the floor of the Atlantic Ocean, radioactive dating of continents and plates, and study both of deep samples of rocks from the continental shelves and of geologic faults.

By the 1960s, a great amount and variety of data were all consistent with the idea that the earth's crust is made up of a few huge, slowly moving plates on which the continents and ocean basins ride. The most difficult argument to overcome—that the surface of the earth is too rigid for continents to move—had proved incorrect. The hot interior of the earth produces a layer of molten rock under the plates, which are moved by convection currents in the layer. In the 1960s, continental drift in the form of the theory of plate tectonics became widely accepted in science and provided geology with a powerful unifying concept.

The theory of plate tectonics was finally accepted because it was supported by the evidence and because it explained so much that had previously seemed obscure or controversial. Such diverse and seemingly unrelated phenomena as earthquakes, volcanoes, the formation of mountain systems and oceans, the shrinking of the Pacific and the widening of the Atlantic, and even some major changes in the earth's climate can now be seen as consequences of the movement of crustal plates.

UNDERSTANDING FIRE

For much of human history, fire was thought to be one of the four basic elements—along with earth, water, and air—out of which everything was made. Burning materials were thought to release the fire that they already contained. Until the eighteenth century, the prevailing scientific theory was that when any object burned, it gave off a substance that carried away weight. This view confirmed what people saw: When a heavy piece of wood was burned, all that was left was a residue of light ashes.

Antoine Lavoisier, a French scientist who made most of his discoveries in the two decades after the American Revolution and was later executed as a victim of the French Revolution, conducted a series of experiments in which he accurately measured all of the substances involved in burning, including the gases used and the gases given off. His measurements demonstrated that the burning process was just the opposite of what people thought. He showed that when substances burn, there is no net gain or loss of weight. When wood burns, for example, the carbon and hydrogen in it combine with oxygen from the air to form water vapor and carbon dioxide, both invisible gases that escape into the air. The total weight of materials produced by burning (gases and ashes) is the same as the total weight of the reacting materials (wood and oxygen).

In unraveling the mystery of burning (a form of combustion), Lavoisier established the modern science of chemistry. Its predecessor, alchemy, had been a search for ways to transform matter—especially to turn lead into gold and to produce an elixir that would confer everlasting life. The search resulted in the accumulation of some descriptive knowledge about materials and processes, but it was unable to lead to an understanding of the nature of materials and how they interact.

Lavoisier invented a whole new enterprise based on a theory of materials, physical laws, and quantitative methods. The intellectual centerpiece of the new science was the concept of the conservation of matter: Combustion and all other chemical processes consist of the interaction of substances such that the total mass of material after the reaction is exactly the same as before it.

For such a radical change, the acceptance of the new chemistry was relatively rapid. One reason was that Lavoisier devised a system for naming substances and for describing their reactions with each other. Being able to make such explicit statements was itself an important step forward, for it encouraged quantitative studies and made possible the widespread dissemination of chemical discoveries without ambiguity. Furthermore, burning came to be seen simply as one example of a category of chemical reactions—oxidation—in which oxygen combines with other elements or compounds and releases energy.

Another reason for the acceptance of the new chemistry was that it fit well with the atomic theory developed by English scientist John Dalton after reading Lavoisier's publications. Dalton elaborated on and refined the ancient Greek ideas of element, compound, atom, and molecule—concepts that Lavoisier had incorporated into his system. This mechanism for developing chemical combinations gave even more specificity to Lavoisier's system of principles. It provided the basis for expressing chemical behavior in quantitative terms.

Thus, for example, when wood burns, each atom of the element carbon combines with two atoms of the element oxygen to form one molecule of the compound carbon dioxide, releasing energy in the process. Flames or high temperatures, however, need not be involved in oxidation reactions. Rusting and the metabolism of sugars in the body are examples of oxidation that occurs at room temperature.

In the three centuries since Lavoisier and Dalton, the system has been vastly extended to account for the configuration taken by atoms when they bond to one another and to describe the inner workings of atoms that account for why they bond as they do.

SPLITTING THE ATOM

A new chapter in our understanding of the structure of matter began at the end of the nineteenth century with the accidental discovery in France that a compound of uranium somehow darkened a wrapped and unexposed photographic plate. Thus began a scientific search for an explanation of this "radioactivity." The pioneer researcher in the new field was Marie Curie, a young Polish-born scientist married to French physicist Pierre Curie. Believing that the radioactivity of uranium-bearing minerals resulted from very small amounts of some highly radioactive substance, Marie Curie attempted, in a series of chemical steps, to produce a pure sample of the substance and to identify it. Her husband put aside his own research to help in the enormous task of separating out an elusive trace from an immense amount of raw material. The result was their discovery of two new elements, both highly radioactive, which they named polonium and radium.

The Curies, who won the Nobel Prize in physics for their research in radioactivity, chose not to exploit their discoveries commercially. In fact, they made radium available to the scientific community so that the nature of radioactivity could be studied further. After Pierre Curie died, Marie Curie continued her research, confident that she could succeed despite the widespread prejudice against women in physical science. She did succeed: She won the 1911 Nobel Prize in chemistry, becoming the first person to win a second Nobel Prize.

Meanwhile, other scientists with better facilities than Marie Curie had available were making major discoveries about radioactivity and proposing bold new theories about it. Ernest Rutherford, a New Zealand-born British physicist, quickly became the leader in this fast-moving field. He and his colleagues discovered that naturally occurring radioactivity in uranium consists of a uranium atom emitting a particle that becomes an atom of the very light element helium, and that what is left behind is no longer a uranium atom but a slightly lighter atom of a different element. Further research indicated that this transmutation was one of a series ending up with a stable isotope of lead. Radium was just one element in the radioactive series.

This transmutation process was a turning point in scientific discovery, for it revealed that atoms are not actually the most basic units of matter; rather, atoms themselves consist of three distinct particles each: a small, massive nucleus—made up of protons and neutrons—surrounded by light electrons. Radioactivity changes the nucleus, whereas chemical reactions affect only the outer electrons.

But the uranium story was far from over. Just before World War II, several German and Austrian scientists showed that when uranium is irradiated by neutrons, isotopes of various elements are produced that have about half the atomic mass of uranium. They were reluctant to accept what now seems the obvious conclusion—that the nucleus of uranium had been induced to split into two roughly equal smaller nuclei. This conclusion was soon proposed by Austrian-born physicist and mathematician Lise Meitner and her nephew Otto Frisch, who introduced the term "fission." They noted, consistent with Einstein's special relativity theory, that if the fission products together had less mass than the original uranium atom, enormous amounts of energy would be released.

Because fission also releases some extra neutrons, which can induce more fissions, it seemed possible that a chain reaction could occur, continually releasing huge amounts of energy. During World War II, a U.S. scientific team led by Italian-born physicist Enrico Fermi demonstrated that if enough uranium were piled together—under carefully controlled conditions—a chain reaction could indeed be sustained. That discovery became the basis of a secret U.S. government project set up to develop nuclear weapons. By the end of the war, the power of an uncontrolled fission reaction had been demonstrated by the explosion of two U.S. fission bombs over Japan. Since the war, fission has continued to be a major component of strategic nuclear weapons developed by several countries, and it has been widely used in the controlled release of energy for transformation into electric power.

EXPLAINING THE DIVERSITY OF LIFE

The intellectual revolution initiated by Darwin sparked great debates. At issue scientifically was how to explain the great diversity of living organisms and of previous organisms evident in the fossil record. The earth was known to be populated with many thousands of different kinds of organisms, and there was abundant evidence that there had once existed many kinds that had become extinct. How did they all get here? Prior to Darwin's time, the prevailing view was that species did not change, that since the beginning of time all known species had been exactly as they were in the present. Perhaps, on rare occasions, an entire species might disappear owing to some catastrophe or by losing out to other species in the competition for food; but no new species could appear.

Nevertheless, in the early nineteenth century, the idea of evolution of species was starting to appear. One line of thought was that organisms would change slightly during their lifetimes in response to environmental conditions, and that those changes could be passed on to their offspring. (One view, for example, was that by stretching to reach leaves high on trees, giraffes—over successive generations—had developed long necks.) Darwin offered a very different mechanism of evolution. He theorized that inherited variations among individuals within a species made some of them more likely than others to survive and have offspring, and that their offspring would inherit those advantages. (Giraffes who had inherited longer necks, therefore, would be more likely to survive and have offspring.) Over successive generations, advantageous characteristics would crowd out others, under some circumstances, and thereby give rise to new species.

Darwin presented his theory, together with a great amount of supporting evidence collected over many years, in a book entitled *Origin of Species,* published in the mid-nineteenth century. Its dramatic effect on biology can be traced to several factors: The argument Darwin presented was sweeping, yet clear and understandable; his line of argument was supported at every point with a wealth of biological and fossil evidence; his comparison of natural selection to the "artificial selection" used in animal breeding was persuasive; and the argument provided a unifying framework for guiding future research.

The scientists who opposed the Darwinian model did so because they disputed some of the mechanisms he proposed for natural selection, or because they believed that it was not predictive in the way Newtonian science was. By the beginning of the twentieth century, however, most biologists had accepted the basic premise that species gradually change, even though the mechanism for biological inheritance was still not altogether understood. Today the debate is no longer about whether evolution occurs but about the details of the mechanisms by which it takes place.

In the general public, there are some people who altogether reject the concept of evolution—not on scientific grounds but on the basis of what they take to be its unacceptable implications: that human beings and other species have common ancestors and are therefore related; that humans and other organisms might have resulted from a process that lacks direction and purpose; and that human beings, like the lower animals, are engaged in a struggle for survival and reproduction. And for some people, the concept of evolution violates the biblical account of the special (and separate) creation of humans and all other species.

At the beginning of the twentieth century, the work of Austrian experimenter Gregor Mendel on inherited characteristics was rediscovered after having passed unnoticed for many years. It held that the traits an organism inherits do not result from a blending of the fluids of the parents but from the transmission of discrete particles—now called genes—from each parent. If organisms have a large number of such particles and some process of random sorting occurs during reproduction, then the variation of individuals within a species—essential for Darwinian evolution—would follow naturally.

Within a quarter of a century of the rediscovery of Mendel's work, discoveries with the microscope showed that genes are organized in strands that split and recombine in ways that furnish each egg or sperm cell with a different combination of genes. By the middle of the twentieth century, genes had been found to be part of DNA molecules, which control the manufacture of the essential materials out of which organisms are made. Study of the chemistry of DNA has brought a dramatic chemical support for biological evolution: The genetic code found in DNA is the same for almost all species of organisms, from bacteria to humans.

DISCOVERING GERMS

Throughout history, people have created explanations for disease. Many diseases have been seen as being spiritual in origin—a punishment for a person's sins or as the capricious behavior of gods or spirits. From ancient times, the most commonly held biological theory was that illness was attributable to some sort of imbalance of body humors (hypothetical fluids that were described by their effects, but not identified chemically). Hence, for thousands of years the treatment of disease consisted of appealing to supernatural powers through offerings, sacrifice, and prayer, or of trying to adjust the body humors by inducing vomiting, bleeding, or purging. However, the introduction of germ theory in the nineteenth century radically changed the explanation of what causes diseases, as well as the nature of their treatment.

As early as the sixteenth century, there was speculation that diseases had natural causes and that the agents of disease were external to the body, and therefore that medical science should consist of identifying those agents and finding chemicals to counteract them. But no one suspected that some of the disease-causing agents might be invisible organisms, since such organisms had not yet been discovered, or even imagined. The improvement of microscope lenses and design in the seventeenth century led to discovery of a vast new world of microscopically small plants and animals, among them bacteria and yeasts. The discovery of those microorganisms, however, did not suggest what effects they might have on humans and other organisms.

The name most closely associated with the germ theory of disease is that of Louis Pasteur, a French chemist. The connection between microorganisms and disease is not immediately apparent—especially since (as we know now) most microorganisms do not cause disease and many are beneficial to us. Pasteur came to the discovery of the role of microorganisms through his studies of what causes milk and wine to spoil. He proved that spoilage and fermentation occur when microorganisms enter them from the air, multiplying rapidly and producing waste products. He showed that food would not spoil if microorganisms were kept out of it or if they were destroyed by heat.

Turning to the study of animal diseases to find practical cures, Pasteur again showed that microorganisms were involved. In the process, he found that infection by disease organisms—germs—caused the body to build up an immunity against subsequent infection by the same organisms, and that it was possible to produce vaccines that would induce the body to build immunity to a disease without actually causing the disease itself. Pasteur did not actually demonstrate rigorously that a particular disease was caused by a particular, identifiable germ; that work was soon accomplished, however, by other scientists.

The consequences of the acceptance of the germ theory of disease were enormous for both science and society. Biologists turned to the identification and investigation of microorganisms, discovering thousands of different bacteria and viruses and gaining a deeper understanding of the interactions between organisms. The practical result was a gradual change in human health practices—the safe handling of food and water; pasteurization of milk; and the use of sanitation measures, quarantine, immunization, and antiseptic surgical procedures—as well as the virtual elimination of some diseases. Today, the modern technology of high-power imaging and biotechnology make it possible to investigate how microorganisms cause disease, how the immune system combats them, and even how they can be manipulated genetically.

HARNESSING POWER

The term "Industrial Revolution" refers to a long period in history during which vast changes occurred in how things were made and in how society was organized. The shift was from a rural handicraft economy to an urban, manufacturing one.

The first changes occurred in the British textile industry in the nineteenth century. Until then, fabrics were made in homes, using essentially the same techniques and tools that had been used for centuries. The machines—like all of the tools of the time—were small, handmade, and powered by muscle, wind, or running water. That picture was radically and irreversibly changed by a series of inventions for spinning and weaving and for using energy resources. Machinery replaced some human crafts; coal replaced humans and animals as the source of power to run machines; and the centralized factory system replaced the distributed, home-centered system of production.

At the heart of the Industrial Revolution was the invention and improvement of the steam engine. A steam engine is a device for changing chemical energy into mechanical work: Fuel is burned, and the heat it gives off is used to turn water into steam, which in turn is used to drive wheels or levers. Steam engines were first developed by inventors in response to the practical need to pump floodwater out of coal and ore mines. After Scottish inventor James Watt greatly improved the steam engine, it also quickly came to be used to drive machines in factories; to move coal in coal mines; and to power railroad locomotives, ships, and later the first automobiles.

The Industrial Revolution happened first in Great Britain—for several reasons: the British inclination to apply scientific knowledge to practical affairs; a political system that favored industrial development; availability of raw materials, especially from the many parts of the British Empire; and the world's greatest merchant fleet, which gave the British access to additional raw materials (such as cotton and wood) and to huge markets for selling textiles. The British also had experienced the introduction of innovations in agriculture, such as cheap plows, which made it possible for fewer workers to produce more food, freeing others to work in the new factories.

The economic and social consequences were profound. Because the new machines of production were expensive, they were accessible mainly to people with large amounts of money, which left out most families. Workshops outside the home that brought workers and machines together resulted in, and grew into, factories—first in textiles and then in other industries. Relatively unskilled workers could tend the new machines, unlike the traditional crafts that required skills learned by long apprenticeship. So surplus farm workers and children could be employed to work for wages.

The Industrial Revolution spread throughout Western Europe and across the Atlantic to North America. Consequently, the nineteenth century was marked in the Western world by increased productivity and the ascendancy of the capitalistic organization of industry. The changes were accompanied by the growth of large, complex, and interrelated industries, and the rapid growth in both total population and a shift from rural to urban areas. There arose a growing tension between, on the one hand, those who controlled and profited from production and, on the other hand, the laborers who worked for wages, which were barely enough to sustain life. To a substantial degree, the major political ideologies of the twentieth century grew out of the economic manifestations of the Industrial Revolution.

In a narrow sense, the Industrial Revolution refers to a particular episode in history. But looked at more broadly, it is far from over. From its beginnings in Great Britain, industrialization spread to some parts of the world much faster than to others, and is only now, reaching some. As it reaches new countries, its economic, political, and social effects have usually been as dramatic as those that occurred in nineteenth-century Europe and North America, but with details shaped by local circumstances.

Moreover, the revolution expanded beyond steam power and the textile industry to incorporate a series of new technological developments, each of which has had its own enormous impact on how people live. In turn, electric, electronic, and computer technologies have radically changed transportation, communications, manufacturing, and health and other technologies; have changed patterns of work and recreation; and have led to greater knowledge of how the world works. (The pace of change in newly industrializing countries may be even greater because the successive waves of innovation arrive more closely spaced in time.) In its own way, each of these continuations of the Industrial Revolution has exhibited the inevitable and growing interdependence of science and technology.

Fragment of hanging, Caucasus (nineteenth–twentieth centuries).

CHAPTER 11
COMMON THEMES

Some important themes pervade science, mathematics, and technology and appear over and over again, whether we are looking at an ancient civilization, the human body, or a comet. They are ideas that transcend disciplinary boundaries and prove fruitful in explanation, in theory, in observation, and in design.

This chapter presents recommendations about some of those ideas and how they apply to science, mathematics, and technology. Here, thematic ideas are presented under six headings: systems, models, stability, patterns of change, evolution, and scale.

RECOMMENDATIONS

SYSTEMS

Any collection of things that have some influence on one another and appear to constitute a unified whole can be thought of as a system. The things can be almost anything, including objects, organisms, machines, processes, ideas, numbers, or organizations. Thinking of a collection of things as a system draws our attention to what needs to be included among the parts to make sense of it, to how its parts interact with one another, and to how the system as a whole relates to other systems. Thinking in terms of systems implies that each part is fully understandable only in relation to the rest of the system.

In defining a system—whether an ecosystem or a solar system, an educational or a monetary system, a physiological or a weather system—we must include enough parts so that their relationship to one another makes some kind of sense. And what makes sense depends on what our purpose is. For example, if we were interested in the energy flow in a forest ecosystem, we would have to include solar input and the decomposition of dead organisms; however, if we were interested only in predator/prey relationships, those could be ignored. If we were interested only in a very rough explanation of the earth's tides, we could neglect all other bodies in the universe except the earth and the moon; however, a more accurate account would require that we also consider the sun as part of the system.

Drawing the boundary of a system well can make the difference between understanding and not understanding what is going on. The conservation of mass during burning, for instance, was not recognized for a long time because the gases produced were not included in the system whose weight was measured. And people believed that maggots could grow spontaneously from garbage until experiments were done in which egg-laying flies were excluded from the system.

Thinking of everything within some boundary as being a system suggests the need to look for certain kinds of influence and behavior. For example, we may consider a system's inputs and outputs. Air and fuel go into an engine; exhaust, heat, and mechanical work come out. Information, sound energy, and electrical energy go into a telephone system; information, sound energy, and heat come out. And we look for what goes into and comes out of any part of the system—the outputs of some parts being inputs for others. For example, the fruit and oxygen that are outputs of plants in an ecosystem are inputs for some animals in the system; the carbon dioxide and droppings that are the output of animals may serve as inputs for the plants.

Part of the output of a system may be fed back to another part. Generally, such feedback serves as a control on what goes on in a system. Feedback can encourage more of what is already happening, discourage it, or modify it to make it something different. For example, some of the amplified sound from a loudspeaker system can feed back into the microphone, then be further amplified, and so on, driving the system to an overload—the familiar feedback squeal. But feedback in a system is not always so prompt. For example, if the deer population in a particular location increases in one year, the greater demand on the scarce winter food supply may result in an increased starvation rate the following year, thus reducing the deer population in that location.

The way that the parts of a system influence one another is not only by transfers of material but also by transfers of information. Such information feedback typically involves a comparison mechanism as part of the system. For example, a thermostat compares the measured temperature in a room to a set value and turns on a heating or cooling device if the difference is too large. Another example is the way in which the leaking of news about government plans before they are officially announced can provoke reactions that cause the plans to be changed; people compare leaked plans to what they would like and then endorse or object to the plans accordingly.

Any part of a system may itself be considered as a system—a subsystem—with its own internal parts and interactions. A deer is both part of an ecosystem and also in itself a system of interacting organs and cells, each of which can also be considered a system. Similarly, any system is likely to be part of a larger system that it influences and that influences it. For example, a state government can be thought of as a system that includes county and city governments as components, but it is itself only one component in a national system of government.

Systems are not mutually exclusive. Systems may be so closely related that there is no way to draw boundaries that separate all parts of one from all parts of the other. Thus, the communication system, the transportation system, and the social system are extensively interrelated; one component—such as an airline pilot—can be a part of all three.

MODELS

A model of something is a simplified imitation of it that we hope can help us understand it better. A model may be a device, a plan, a drawing, an equation, a computer program, or even just a mental image. Whether models are physical, mathematical, or conceptual, their value lies in suggesting how things either do work or might work. For example, once the heart has been likened to a pump to explain what it does, the inference may be made that the engineering principles used in designing pumps could be helpful in understanding heart disease. When a model does not mimic the phenomenon well, the nature of the discrepancy is a clue to how the model can be improved. Models may also mislead, however, suggesting characteristics that are not really shared with what is being modeled. Fire was long taken as a model of energy transformation in the sun, for example, but nothing in the sun turned out to be burning.

Physical Models

The most familiar meaning of the term "model" is the physical model—an actual device or process that behaves enough like the phenomenon being modeled that we can hope to learn something from it. Typically, a physical model is easier to work with than what it represents because it is smaller in size, less expensive in terms of materials, or shorter in duration.

Experiments in which variables are closely controlled can be done on a physical model in the hope that its response will be like that of the full-scale phenomenon. For example, a scale model of an airplane can be used in a wind tunnel to investigate the effects of different wing shapes. Human biological processes can be modeled by using laboratory animals or cultures in test tubes to test medical treatments for possible use on people. Social processes too can be modeled, as when a new method of instruction is tried out in a single classroom rather than in a whole school system. But the scaling need not always be toward smaller and cheaper. Microscopic phenomena such as molecular configurations may require much larger models that can be measured and manipulated by hand.

A model can be scaled in time as well as in size and materials. Something may take so inconveniently long to occur that we observe only a segment of it. For example, we may want to know what people will remember years later of what they have been taught in a school course, but we settle for testing them only a week later. Short-run models may attempt to compress long-term effects by increasing the rates at which events occur. One example is genetic experimentation on organisms such as bacteria, flies, and mice that have large numbers of generations in a relatively short time span. Another important example is giving massive doses of chemicals to laboratory animals to try to get in a short time the effect that smaller doses would produce over a long time. A mechanical example is the destructive testing of products, using machines to simulate in hours the wear on, say, shoes or weapons that would occur over years in normal use. On the other hand, very rapid phenomena may require slowed-down models, such as slow-motion depiction of the motion of birds, dancers, or colliding cars.

The behavior of a physical model cannot be expected ever to represent the full-scale phenomenon with complete accuracy, not even in the limited set of characteristics being studied. If a model boat is very small, the way water flows past it will be significantly different from a real ocean and boat; if only one class in a school uses a new method, the specialness of it may make it more successful than the method would be if it were commonplace; large doses of a drug may have different kinds of effects (even killing instead of curing), not just quicker effects. The inappropriateness of a model may be related to such factors as changes in scale or the presence of qualitative differences that are not taken into account in the model (for example, rats may be sensitive to drugs that people are not, and vice versa).

Conceptual Models

One way to give an unfamiliar thing meaning is to liken it to some familiar thing—that is, to use metaphor or analogy. Thus, automobiles were first called horseless carriages. Living "cells" were so called because in plants they seemed to be lined up in rows like rooms in a monastery; an electric "current" was an analogy to a flow of water; the electrons in atoms were said to be arranged around the nucleus in "shells." In each case, the metaphor or analogy is based on some attributes of similarity—but only some. Living cells do not have doors; electric currents are not wet; and electron shells do not have hard surfaces. So we can be misled, as well as assisted, by metaphor or analogy, depending on whether inappropriate aspects of likeness are inferred along with the appropriate aspects. For example, the metaphor for the repeated branching of species in the "tree of evolution" may incline one to think not just of branching but also of upward progress; the metaphor of a bush, on the other hand, suggests that the branching of evolution produces great diversity in all directions, without a preferred direction that constitutes progress. If some phenomenon is very unlike our ordinary experience, such as quantum phenomena on an atomic scale, there may be no single familiar thing to which we can liken it.

Like any model, a conceptual model may have only limited usefulness. On the one hand, it may be too simple. For example, it is useful to think of molecules of a gas as tiny elastic balls that are endlessly moving about, bouncing off one another; to accommodate other phenomena, however, such a model has to be greatly modified to include moving parts within each ball. On the other hand, a model may be too complex for practical use. The accuracy of models of complex systems such as global population, weather, and social integration is limited by the large number of interacting variables that need to be dealt with simultaneously. Or, an abstract model may fit observations very well, but have no intuitive meaning. In modeling the behavior of molecules, for instance, we have to rely on a mathematical description that may not evoke any associated mental picture. Any model may have some irrelevant features that intrude on our use of it. For example, because of their high visibility and status, athletes and entertainers may be taken as role models by children not only in the aspects in which they excel but also in irrelevant—and perhaps distinctly less than ideal—aspects.

Mathematical Models

The basic idea of mathematical modeling is to find a mathematical relationship that behaves in the same way the system of interest does. (The system in this case can be other abstractions, as well as physical or biological phenomena.) For example, the increasing speed of a falling rock can be represented by the symbolic relation $v = gt$, where g has a fixed value. The model implies that the speed of fall (v) increases in proportion to the time of fall (t). A mathematical model makes it possible to predict what phenomena may be like in situations outside of those in which they have already been observed—but only what they *may* be like. Often, it is fairly easy to find a mathematical model that fits a phenomenon over a small range of conditions (such as temperature or time), but it may not fit well over a wider range. Although $v = gt$ does apply accurately to objects such as rocks falling (from rest) more than a few meters, it does not fit the phenomenon well if the object is a leaf (air drag limits its speed) or if the fall is a much larger distance (the drag increases, the force of gravity changes).

Mathematical models may include a set of rules and instructions that specifies precisely a series of steps to be taken, whether the steps are arithmetic, logical, or geometric. Sometimes even very simple rules and instructions can have consequences that are extremely difficult to predict without actually carrying out the steps. High-speed computers can explore what the consequences would be of carrying out very long or complicated instructions. For example, a nuclear power station can be designed to have detectors and alarms in all parts of the control system, but predicting what would happen under various complex circumstances can be very difficult. The mathematical models for all parts of the control system can be linked together to simulate how the system would operate under various conditions of failure.

What kind of model is most appropriate varies with the situation. If the underlying principles are poorly understood, or if the mathematics of known principles is very complicated, a physical model may be preferable; such has been the case, for example, with the turbulent flow of fluids. The increasing computational speed of computers makes mathematical modeling and the resulting graphic simulation suitable for more and more kinds of problems.

CONSTANCY

In science, mathematics, and engineering, there is great interest in ways in which systems do not change. In trying to understand systems, we look for simplifying principles, and aspects of systems that do not change are clearly simplifying. In designing systems, we often want to ensure that some characteristics of them remain predictably the same.

Stability and Equilibrium

The ultimate fate of most physical systems, as energy available for action dissipates, is that they settle into a steady state, or state of equilibrium. For example, a falling rock comes to rest at the foot of a cliff, or a glass of ice water melts and warms up to room temperature. In these states, all forces are balanced, all processes of change appear to have stopped—and will remain so until something new is done to the system, after which it will eventually settle into a new equilibrium. If a new ice cube is added to the glass of water, heat from the environment will transfer into the glass until the glass again contains water at room temperature. If a consumer product with a stable price is improved, the price may rise until the expense causes the number of buyers to level off at some higher equilibrium value.

The idea of equilibrium can also be applied to systems in which there is continual change going on, as long as the changes counterbalance one another. For example, the job market can be thought of as being in equilibrium if the total number of unemployed people stays very nearly the same, even though many people are losing jobs and others are being hired. Or an ecosystem is in equilibrium if members of every species are dying at the same rate at which they are being reproduced.

From a molecular viewpoint, all steady states belie a continual activity of molecules. For example, when a bottle of club soda is capped, molecules of water and carbon dioxide escaping from the solution into the air above increase in concentration until the rate of return to the liquid is as great as the rate of escape. The escaping and returning continue at a high rate, while the observable concentrations and pressures remain in a steady state indefinitely. In the liquid itself, some molecules of water and carbon dioxide are always combining and others are coming apart, thereby maintaining an equilibrium concentration of the mild acid that gives the tingly taste.

Some processes, however, are not so readily reversible. If the products of a chemical combination do not readily separate again, or if a gas evaporating out of solution drifts away, then the process will continue in one direction until no reactants are left—leaving a static rather than dynamic equilibrium. Also, a system can be in a condition that is stable over small disturbances but not large ones. For example, a falling stone that comes to rest part way down a hill will stay there if only small forces disturb it; but a good kick may free it to start downhill again toward a more stable condition at the bottom.

Many systems include feedback subsystems that serve to keep some aspect of the system constant, or at least within specified limits of variation. A thermostat designed to regulate a heating or cooling system is a common example, as is the set of biological reactions in mammals that keeps their body temperatures within a narrow range. Such mechanisms may fail, however, if conditions go far outside their usual range of operation (that is what happens, for example, when sunstroke shuts down the human body's cooling system).

Conservation

Some aspects of systems have the remarkable property of always being conserved. If the quantity is reduced in one place, an exactly equal increase always shows up somewhere else. If a system is closed to such a quantity, with none entering or leaving its boundaries, then the total amount inside will not change, no matter how much the system may change in other ways. Whatever happens inside the system—parts dissolving, exploding, decaying, or changing in any way—there are some total quantities that remain precisely the same. In an explosion of a charge of dynamite, for example, the total mass, momentum, and energy of all the products (including fragments, gases, heat, and light) remains constant.

Symmetry

Besides the constancy of totals, there are constancies in form. A Ping-Pong ball looks very much the same no matter how it is turned. An egg, on the other hand, will continue to look the same if it is turned around its long axis, but not if it is turned any other way. A human face looks very different if it is turned upside down, but not if it is reflected left for right, as in a mirror. The outline of an octagonal stop sign or a starfish will look the same after being turned through a particular angle. Natural symmetry of form often indicates symmetrical processes of development. Clay bowls, for example, are symmetrical because they were continually rotated while being formed by steady hands. Almost all land animals are approximately left-and-right symmetrical, which can be traced back to a symmetrical distribution of cells in the early embryo.

But symmetry is not a matter of geometry only. Operations on numbers and symbols can also show invariance. The simplest may be that exchanging the terms in the sum $X + Y$ results in the same value: $Y + X = X + Y$. But $X - Y$ shows a different kind of symmetry: $Y - X$ is the negative of $X - Y$. In higher mathematics, there can be some very subtle kinds of symmetry. Because mathematics is so widely used to model how things in the world behave, symmetries in mathematics can suggest unsuspected symmetries that underlie physical phenomena.

PATTERNS OF CHANGE

Patterns of change are of special interest in the sciences: Descriptions of change are important for predicting what will happen; analysis of change is essential for understanding what is going on, as well as for predicting what will happen; and control of change is essential for the design of technological systems. We can distinguish three general categories: (1) changes that are steady trends, (2) changes that occur in cycles, and (3) changes that are irregular. A system may have all three kinds of change occurring together.

Trends

Steady changes are found in many phenomena, from the increasing speed of a falling rock, to the mutation of genes in a population, to the decay of radioactive nuclei. Not all of these trends are steady in the same sense, but all progress in one direction and have fairly simple mathematical descriptions. The rate of radioactive decay in a sample of rock diminishes with time, but it is a constant proportion of the number of undecayed nuclei left. Progressive changes that fit an identifiable mathematical form can be used to estimate how long a process has been going on. For example, the remaining radioactivity of rocks indicates how long ago they were formed, and the current number of differences in the DNA of two species may indicate how many generations ago they had a common ancestor.

Cycles

A sequence of changes that happens over and over again—a cyclic change—is also familiar in many phenomena, such as the seasonal cycles of weather, the vibration of a guitar string, body temperature in mammals, and the sweep of an electron beam across a television tube. Cycles are characterized by how large the range of variation is from maximum to minimum, by how long a cycle takes, and by exactly when its peaks occur. For the daily cycle in human body temperature, for example, the variation is about a degree; the cycle repeats about every 24 hours; and the peaks usually occur in late afternoon. For the guitar string, the variation in movement is about a millimeter, and each cycle takes about a thousandth of a second. Cycles can be as long as the thousands of years between ice ages, or shorter than a billionth of a second in electric oscillators. Many phenomena, such as earthquakes and ice ages, have patterns of change that are persistent in form but irregular in their period—we know that it is their nature to recur, but we cannot predict precisely when.

The extent of variation during a cycle can be so great as to disrupt the system, such as when vibrations crumble buildings in earthquakes, or can be almost too small to detect, apparently lost in the random activity of the system. What is random and what is regular, however, is not always obvious from merely looking at data. Data that appear to be completely irregular may be shown by statistical analysis to have underlying trends or cycles. On the other hand, trends or cycles that appear in data may sometimes be shown by statistical analysis to be easily explainable as being attributable only to randomness or coincidence.

Cyclic change is commonly found when there are feedback effects in a system; as, for example, when a change in any direction gives rise to forces that oppose the change. A system with such feedback that acts slowly is likely to change a significant amount before it is nudged back toward normal; and when it gets back to normal, the momentum of change may carry it some distance in the opposite direction, and so on, producing a more or less regular cycle. Biological systems as small as single cells have chemical cycles that result from feedback because the products of reactions affect the rates at which the reactions occur. In complex organisms, the feedback effects of neural and hormonal control systems on one another produce distinct rhythms in many body functions (for example, in blood cell counts, in sensitivity to drugs, in alertness, and even in mood). On the level of human society, any trend is eventually likely to evoke reactions that oppose it, so there are many social cycles: The swing of the pendulum is evident in everything from economics to fashions to the philosophy of education.

Chaos

On a sufficiently small scale, all change appears to have a random component. Just which reactions occur and when they occur in and among atoms appear to be unpredictable even in principle. Moreover, a random appearance may sometimes result entirely from an underlying regularity. Even some fairly simple and precisely defined processes, when repeated many times, may have immensely complicated, apparently chaotic results. Most systems above the molecular scale involve the interactions of so many parts and forces and are so sensitive to tiny differences in conditions that their precise behavior is unpredictable, appearing to be random, at least in part.

In spite of the unpredictability of details, however, the summary behavior of some large systems may be highly predictable. Changes in the pressure and temperature of a gas in equilibrium can often be predicted with great accuracy, despite the chaotic motion of its molecules and the scientist's inability to predict the motion of any one molecule. The average distribution of leaves around a tree or the percentage of heads in a long series of coin tosses will occur with predictable reliability from one occasion to the next. Likewise, prediction of an individual's behavior is likely to be less reliable than prediction of the average behavior of members of a group of individuals.

Many systems also show approximate stability in cyclic behavior. They go through pretty much the same sequence of states over and over, although the details are never quite the same twice: for example, the orbiting of the moon around the earth, the human cycles of sleep and wakefulness, and the cyclic fluctuations in populations of predator and prey. Although such systems involve interplay of highly complex influences, they may persist indefinitely in approximating a single, very simple cycle. Small disturbances will result only in a return to the same approximate cycle, although large disturbances may send the system into very different behavior—which may, however, be another simple cycle.

EVOLUTION

The general idea of evolution, which dates back at least to ancient Greece, is that the present arises from the materials and forms of the past, more or less gradually, and in explicable ways. This is how the solar system, the face of the earth, and life forms on earth have evolved from their earliest states and continue to evolve. The idea of evolution applies also, although perhaps more loosely, to language, literature, music, political parties, nations, science, mathematics, and technological design. Each new development in these human endeavors has grown out of the forms that preceded them, and those earlier forms themselves had evolved from still earlier forms.

Possibilities

The fins of whales, the wings of bats, the hands of people, and the paws of cats all appear to have evolved from the same set of bones in the feet of ancient reptilian ancestors. The genetic instructions for the set of bones were there, and the fins or paws resulted from natural selection of changes in those instructions over many generations. A fully formed eye did not appear all of a sudden where there was no light-sensitive organ before; nor did an automobile appear where there had been no four-wheeled vehicle; nor did the theory of gravitation arise until after generations of thought about forces and orbits.

What can happen next is limited to some extent by what has happened so far. But how limited is it? An extreme view is that what happens next is completely determined by what has happened so far—there is only one possible future. There are two somewhat different reasons for doubting this view. One is that many processes are chaotic—even vanishingly small differences in conditions can produce large differences in outcomes. The other reason is that there is a completely unpredictable random factor in the behavior of atoms, which act at the base of things. So it seems that the present limits the possibilities for what happens next, but does not completely determine it.

Rates

Usually, people think of evolution of a system as proceeding gradually, with a series of intermediate states between the old and the new. This does not mean that evolutionary change is necessarily slow. Intermediate stages may occur very rapidly and even be difficult to identify. Explosions, for example, involve a succession of changes that occur almost too rapidly to track—whether the explosions are electric as in lightning, chemical as in automobile engines, or nuclear as in stars. What is too rapid, however, depends on how finely the data can be separated in time. Consider, for example, a collection of fossils of fairly rare organisms known to have existed in a period that lasted many thousands of years. In this case, evolutionary changes that occurred within a thousand years would be impossible to track precisely. And some evolutionary change does occur in jumps. For instance, new biological developments do not arise only by successive rearrangement of existing genes but sometimes by the abrupt mutation of a gene into a new form. On an atomic scale, electrons change from one energy state to another with no possible intermediate states. For both the gene and the electron, however, the new situation is limited by, and explicable from, the previous one.

Interactions

Evolution does not occur in isolation. While one life form is evolving, so are others around it. While a line of political thought is evolving, so are the political conditions around it. And more generally, the environment to which things and ideas must respond changes even as they are evolving—perhaps impeding, or perhaps facilitating, their change in a particular direction. For example, abrupt changes in a long-steady climate may lead to extinction of species that have become well adapted to it. The economic hardships in Europe after World War I facilitated the rise of the Fascists who instigated World War II. The availability of recently developed mathematical ideas of curved spaces enabled Einstein to put his ideas of relativity into a convincing quantitative form. The development of electricity fostered the spread of rapid long-distance communications.

SCALE

The ranges of magnitudes in our universe—sizes, durations, speeds, and so on—are immense. Many of the discoveries of physical science are virtually incomprehensible to us because they involve phenomena on scales far removed from human experience. We can measure, say, the speed of light, the distance to the nearest stars, the number of stars in the galaxy, and the age of the sun, but these magnitudes are far greater than we can comprehend intuitively. In the other direction, we can determine the size of atoms, their vast numbers, and how quickly interactions among them occur, but these extremes also exceed our powers of intuitive comprehension. Our limited perceptions and information-processing ca-

pacities simply cannot handle the whole range. Nevertheless, we can represent such magnitudes in abstract mathematical terms (for example, billions of billions) and seek relationships among them that make sense.

Large changes in scale typically are accompanied by changes in the kind of phenomena that occur. For instance, on a familiar human scale, a small puff of gas emitted from an orbiting satellite dissipates into space; on an astronomical scale, a gas cloud in space with enough mass is pulled together by mutual gravitational forces into a hot ball that ignites nuclear fusion and becomes a star. On a human scale, substances and energy are endlessly divisible; on an atomic scale, matter cannot be divided and still keep its identity, and energy can change only by discrete jumps. The distance around a tree is much greater for a small insect than for a squirrel—in that on the scale of the insect's size there are many hills and valleys to traverse, whereas for the squirrel there are none.

Even within realms of space and time that are directly familiar to us, scale plays an important role. Buildings, animals, and social organizations cannot be made significantly larger or smaller without experiencing fundamental changes in their structure or behavior. For example, it is not possible to make a forty-story building of precisely the same design and materials commonly used for a four-story building because (among other things) it would collapse under its own weight. As objects increase in size, their volume increases faster than their surface area. Properties that depend on volume, such as capacity and weight, therefore change out of proportion to properties that depend on area, such as strength of supports or surface activity. For example, a substance dissolves much more quickly when it is finely ground than when it is in a lump because the ratio of surface area to volume is much greater. A microorganism can exchange substances with its environment directly through its surface, whereas a larger organism requires specialized, highly branched surfaces (such as in lungs, blood vessels, and roots).

Internal connections also show a strong scale effect. The number of possible pairs of things (for example, friendships or telephone connections) increases approximately as the square of the number of things. Thus, a community ten times as large will have approximately a hundred times as many possible telephone connections between residents. More generally, a city is not simply a large village, since almost everything that characterizes a city—services, work patterns, methods of governance—is necessarily different from, not just larger than, that in a village. Systems sometimes include so many interconnected components that they defy precise description. As the scale of complexity increases, eventually we must resort to summary characteristics, such as averages over very large numbers of atoms or instants of time, or descriptions of typical examples.

Systems of sufficient complexity may show characteristics that are not predictable from the interaction of their components, even when those interactions are well understood. In such circumstances, principles that do not make direct reference to the underlying mechanisms—but that are not inconsistent with them—may be required. For example, the process of scouring by glaciers can be referred to in geology without reference to the underlying physics of electric forces and crystal structure of minerals in rocks; we can think of the heart in terms of the volume of blood it delivers, regardless of how its cells behave; we can predict the likely response of someone to a message without reference to how brain cells function; or we can analyze the effects of pressure groups in politics without necessarily referring to any particular people. Such phenomena can be understood at various levels of complexity, even though the full explanation of such things is often reduced to a scale far outside our direct experience.

Page from a Leonardo da Vinci notebook (fifteenth–sixteenth centuries).

CHAPTER 12

HABITS OF MIND

Throughout history, people have concerned themselves with the transmission of shared values, attitudes, and skills from one generation to the next. All three were taught long before formal schooling was invented. Even today, it is evident that family, religion, peers, books, news and entertainment media, and general life experiences are the chief influences in shaping people's views of knowledge, learning, and other aspects of life. Science, mathematics, and technology—in the context of schooling—can also play a key role in the process, for they are built upon a distinctive set of values, they reflect and respond to the values of society generally, and they are increasingly influential in shaping shared cultural values. Thus, to the degree that schooling concerns itself with values and attitudes—a matter of great sensitivity in a society that prizes cultural diversity and individuality and is wary of ideology—it must take scientific values and attitudes into account when preparing young people for life beyond school.

Similarly, there are certain thinking skills associated with science, mathematics, and technology that young people need to develop during their school years. These are mostly, but not exclusively, mathematical and logical skills that are essential tools for both formal and informal learning and for a lifetime of participation in society as a whole.

Taken together, these values, attitudes, and skills can be thought of as habits of mind because they all relate directly to a person's outlook on knowledge and learning and ways of thinking and acting.

This chapter presents recommendations about values, attitudes, and skills in the context of science education. The first part of the chapter focuses on four specific aspects of values and attitudes: the values inherent in science, mathematics, and technology; the social value of science and technology; the reinforcement of general social values; and people's attitudes toward their own ability to understand science and mathematics. The second part of the chapter focuses on six specific aspects of skills that emphasize basic mathematics, critical thinking, and communication.

RECOMMENDATIONS

VALUES AND ATTITUDES

Science education should contribute to people's knowledge of the shared values of scientists, mathematicians, and engineers; reinforcement of general societal values; the inculcation in people of informed, balanced beliefs about the social value of science, mathematics, and technology; and the development in young people of positive attitudes toward learning science, mathematics, and technology.

Knowledge of the Values Inherent in Science, Mathematics, and Technology

Science, mathematics, and technology incorporate particular values, some of which are different in kind or intensity from those in other human enterprises, such as business, law, and the arts. To understand science, mathematics, and technology, it is essential to be aware of some of the values that underlie them and give them character and that are shared by the people who work in the three fields. These values are evident in the recommendations presented in this report's three chapters on the nature of science, mathematics, and technology, which consider the importance of the following verifiable data, testable hypotheses, and predictability in science; of rigorous proof and elegance in mathematics; and of optimum design in technology.

Reinforcement of General Societal Values

Culturally, science can be viewed as both revolutionary and conservative. The knowledge it generates sometimes forces us to change—even discard—beliefs we have long held about ourselves and our significance in the grand scheme of things. The revolutions that we associate with Newton, Darwin, and Lyell have had as much to do with our sense of humanity as they do with our knowledge of the earth and its inhabitants. Moreover, scientific knowledge can surprise us, even trouble us, especially when we discover that our world is not as we perceive it or would like it to be. The discovery that the earth is billions, rather than thousands, of years old may be a case in point. Such discoveries can be so distressing that it may take us years—or perhaps take society as a whole several generations—to come to terms with the new knowledge. Part of the price we pay for obtaining knowledge is that it may make us uncomfortable, at least initially. Becoming aware of the impact of scientific and technological developments on human beliefs and feelings should be part of everyone's science education.

It is also important for people to be aware that science is based upon everyday values even as it questions our understanding of the world and ourselves. Indeed, science is in many respects the systematic application of some highly regarded human values—integrity, diligence, fairness, curiosity, openness to new ideas, skepticism, and imagination. Scientists did not invent any of these values, and they are not the only people who hold them. But the broad field of science does incorporate and emphasize such values and dramatically demonstrates just how important they are for advancing human knowledge and welfare. Therefore, if science is taught effectively, the result will be to reinforce such generally desirable human attitudes and values.

Science education is in a particularly strong position to foster three of these attitudes and values—curiosity, openness to new ideas, and skepticism.

Curiosity. Scientists thrive on curiosity— and so do children. Children enter school alive with questions about everything in sight, and they differ from scientists only in not yet having learned how to go about finding answers and checking to see how good those answers are. Science education that fosters curiosity and teaches children how to channel that curiosity in productive ways serves both students and society well.

Openness to New Ideas. New ideas are essential for the growth of science—and for human activities in general. People with closed minds miss the joy of discovery and the satisfaction of intellectual growth throughout life. Because, as this report makes clear, the purpose of science education is not exclusively to produce scientists, it should help all students understand the great importance of carefully considering ideas that at first may seem disquieting to them or at odds with what they generally believe. The competition among ideas is a major source of tensions within science, between science and society, and within society. Science education should document the nature of such tensions from the history of science—and it should help students see the value to themselves and society of participating in the push and pull of conflicting ideas.

Skepticism. Science is characterized as much by skepticism as by openness. Although a new theory may receive serious attention, it rarely gains widespread acceptance in science until its advocates can show that it is borne out by the evidence, is logically consistent with other principles that are not in question, explains more than its rival theories, and has the potential to lead to new knowledge. Because most scientists are skeptical about all new theories, such acceptance is usually a process of verification and refutation that can take years or even decades to run its course. Science education can help students to see the social value of systematic skepticism and to develop a healthy balance in their own minds between openness and skepticism.

The Social Value of Science, Mathematics, and Technology

There is another sense in which values come into play in thinking about the outcomes of the learning process. Quite apart from what scientific values students may adopt for themselves, there is the issue of what students should know and believe about the general social value of those endeavors. Is it necessary that every graduate be convinced of the great value to society of science, mathematics, and technology?

On balance, science, mathematics, and technology have advanced the quality of human existence, and students should become thoughtful supporters of them. But since science itself esteems independent thought so highly, it follows that teachers should not attempt to simply indoctrinate students into becoming uncritical supporters of science. Rather, they should take the position that in achieving the goals recommended in this report, students will end up with balanced views of the value of science, mathematics, and technology, and not be either uncritically positive or antagonistic.

Attitudes Toward Learning Science, Mathematics, and Technology

Students in elementary school have a spontaneous interest in nature and numbers. Nevertheless, many students emerge from school fearing mathematics and disdaining science as too dull and too hard to learn. They see science only as an academic activity, not as a way of understanding the world in which they live. The consequences of this aversion are severe, for it means that the lives of too many students are being limited and the nation's overall pool of talent from which scientists, mathematicians, and engineers are drawn is smaller than it should be.

The schools may not be able to turn this situation around by themselves, but they are essential to any realistic hope of doing so. It is within teachers' power to foster positive attitudes among their students. If they choose significant, accessible, and exciting topics in science and mathematics, if they feature teamwork as well as competition among students, if they focus on exploring and understanding more than the rote memorization of terms, and if they make sure all their students know they are expected to explore and learn and have their achievements acknowledged, then nearly all of those students will indeed learn. And in learning successfully, students will learn the most important lesson of all—namely, that they are able to do so.

SKILLS

The recommendations presented in the preceding chapters are mostly about knowledge. However, they also imply that knowledge should be understood in ways that will enable it to be used in solving problems. In this sense, all of the foregoing recommendations are about thinking skills. Putting this the other way around, students are likely to learn thinking skills only in the process of coming to understand something substantive about the world, of encountering them in many different contexts and situations, and of using them repeatedly.

Computational Skills

Repeated experience with computations in meaningful contexts will also foster the higher-level skill of judging when computations can most appropriately be made in one's head or on paper, or with the help of a calculator or computer. Each of these methods has a legitimate role in problem solving, although their roles may be different in different circumstances.

Basic Number Skills. In everyday life, one must be able to make simple calculations in one's mind. However, the actual amount of mental arithmetic needed is quite limited and is well within the ability of all normal individuals to learn. This skill requires, first of all, that the individual memorize and be able to recall immediately certain number facts:

- The sums, differences, and products of whole numbers from 1 to 10.

- The decimal equivalents of key common fractions—halves, thirds, two-thirds, fourths, three-fourths, fifths, tenths, and hundredths (but not sixths, sevenths, ninths, and other fractions rarely encountered by most people).

- The relation between decimal fractions and percentages (such as the equivalence of 0.23 and 23 percent).

- The relations among 10, 100, 1,000, 1 million, and 1 billion (for example, knowing that 1 million is a thousand thousands). Expressed as powers of 10, these relations are, successively, 10^1, 10^2, 10^3, 10^6, and 10^9.

There are two kinds of mental calculations that everyone ought to be able to perform:

- The addition of any two numbers that have two digits each.

- The multiplication and division of any number by 2, 10, and 100, to one or two significant digits.

Calculator Skills. In everyday life, and especially in the workplace, almost everyone encounters the need to make calculations. Until recently, paper and pencil were the most common means of solving problems that people could not do by mental arithmetic. For most students, school mathematics has meant doing calculations on paper. This usually takes the form of learning how to do long division, find percentages, or compute ratios, but not of learning why those algorithms work, when to use them, or how to make sense out of the answers.

The advent of the small, inexpensive electronic calculator has made it possible to change that situation radically. Because calculators are so fast, they can make available instructional time in school for doing and learning real mathematics. Students can readily learn how to figure out steps for solving ordinary numerical problems, which operations to use, and how to check the reasonableness of their answers. Universal numeracy becomes a real possibility.

The advantage of the calculator is not only pedagogical. Paper-and-pencil calculations are slow, prone to error, and as conceptually mysterious to most users as electronic tools are. When precision is desired, when the numbers being dealt with have multiple digits, or when the computation has several steps, the calculator offers many practical advantages over paper and pencil. But those advantages cannot be realized unless people learn how to use calculators intelligently. Calculator use does require skill, does not compensate for human errors of reasoning, often delivers answers with more precision than the data merit, and can be undermined by operator error. The key is for students to start using calculators early and to use them throughout the school years in as many different subjects as possible.

Everyone should be able to use a calculator to do the following:

- Add, subtract, multiply, and divide any two whole or decimal numbers (but not powers, roots, or trigonometric functions).

- Find the decimal equivalent of any fraction.

- Calculate what percentage one number is of another and take a percentage of any number (for example, 10 percent off, 60 percent gain).

- Find the reciprocal of any number.

- Determine rates from magnitudes (for example, speed from time and distance) and magnitudes from rates (for example, the amount of simple interest to be paid on the basis of knowing the interest rate and the principal, but not calculations using compound interest).

- Calculate circumferences and areas of rectangles, triangles, and circles, and the volumes of rectangular solids.

- Find the mean of a set of data.

- Determine by numerical substitution the value of simple algebraic expressions— for example, the expressions $aX + bY$, $a(A + B)$, and $(A - B)/(C + D)$.

- Convert compound units (such as yen per dollar into dollars per yen, miles per hour into feet per second).

To make full and effective use of calculators, everyone should also be able to do the following:

• Read and follow step-by-step instructions given in calculator manuals when learning new procedures.

• Make up and write out simple algorithms for solving problems that take several steps.

• Figure out what the unit (such as seconds, square inches, or dollars per tankful) of the answer will be from the inputs to the calculation. Most real-world calculations have to do with magnitudes (numbers associated with units), but ordinary calculators only respond with numbers. The user must be able to translate the calculator's "57," for example, into "57 miles per hour."

• Round off the number appearing in the calculator answer to a number of significant figures that accurately reflects those of the inputs. For example, for the speed of a car that goes 200 kilometers (give or take a kilometer or two) in 3 hours (give or take a minute or two), 67 kilometers per hour is probably accurate enough, 66.67 kilometers per hour is clearly going too far, and 66.666667 kilometers per hour is ridiculous.

• Judge whether an answer is reasonable by comparing it to an estimated answer. A result of 6.7 kilometers per hour or 667 kilometers per hour for the highway speed of an automobile, for example, should be rejected on sight.

Estimation Skills. There are many circumstances in which an approximate answer is as useful as a precise one. Indeed, this may be the rule rather than the exception. Estimating approximate answers can often take the place of making a precise measurement or a careful calculation but in most cases will serve as a check of calculations made using electronic calculators or paper and pencil. Skill in estimation is based on a sense of what an adequate degree of precision is in a particular situation, which in turn depends on understanding the context of the problem and the purpose of the calculation. Among particular estimation skills, everyone should be able to estimate the following:

• Familiar lengths and weights, and also time periods.

• Distances and travel times from maps.

• The actual sizes of objects, based on the use of scale drawings.

• Probabilities of outcomes of familiar situations, either on the basis of history (such as the fact that a certain football team has won its opening game eight times in the last 10 years) or on the basis of the number of possible outcomes (for example, there are six sides on a die).

It often happens that an answer shown on a calculator is wrong because the information entered was wrong, the information was entered incorrectly, or the wrong sequence of operations was used. In situations where there is no basis for judging the appropriateness of an answer arrived at by calculation, everyone should be able to figure out a rough estimate of what the answer ought to be before accepting it. This involves being able to do three things:

• Carry out rough estimates of sums, differences, products, quotients, fractions, and percentages.

• Trace the source of any large disparity between the estimate and the calculated answer.

• Specify a magnitude only to the nearest power of 10. Thus, the population of the world is "on the order" of 10^9 (a billion) or 10^{10} (ten billion). Something that is improved by "an order of magnitude" changes by a factor of about 10—that is, anything from 4 or 5 times to 20 or 30 times larger (or smaller). A factor of 40 or a few hundred, for instance, would be more like two orders of magnitude.

Manipulation and Observation Skills

Everyone should acquire the ability to handle common materials and tools for dealing with household and other everyday technologies, for making careful observations, and for handling information. These include being able to do the following:

• Keep a notebook that accurately describes observations made, that carefully distinguishes actual observations from ideas and speculations about what was observed, and that is understandable weeks or months later.

• Store and retrieve computer information using topical, alphabetical, numerical, and key-word files, and use simple files of the individual's own devising.

• Enter and retrieve information on a computer, using standard software.

• Use appropriate instruments to make direct measurements of length, volume, weight, time interval, and temperature. Besides selecting the right instrument, this skill entails using a precision relevant to the situation (for example, measuring to the nearest quarter-inch is not good enough for making a cabinet, but is better than what is needed for building a long fence).

• Take readings from standard meter displays, both analog and digital, and make prescribed settings on dials, meters, and switches.

• Make electrical connections with various plugs and sockets and screw terminals, exercising reasonable safety.

• Shape, fasten, and unfasten common materials (such as wood, clay, paper, plastics, and metal) using ordinary hand and power tools, exercising reasonable safety.

• Dilute and mix dry and liquid materials (in the kitchen, garage, or laboratory) in prescribed proportions, exercising reasonable safety.

• Do simple troubleshooting on common mechanical and electrical systems, identifying and eliminating some possible causes of malfunction (such as a burned-out bulb versus an unplugged cord versus a faulty cord or switch in a house, or an empty gas tank versus a run-down battery versus a flooded carburetor in an automobile).

• Compare consumer products on the basis of basic features, performance, durability, and cost, making supportable personal trade-offs.

• Look for the implications of changes in one part of a system—inputs, outputs, or connections—for the operation of other parts.

Communication Skills

Discourse in science, mathematics, and technology calls for the ability to communicate ideas and share information with fidelity and clarity, and to read and listen with understanding. Some of the skills involved are specific to science, mathematics, and technology, and others are general—although even those are not independent of content. Everyone should have the skills that will enable him or her to do the following:

• Express orally and in writing the basic ideas covered by the recommendations in this report. This requires, above all, that students acquire some understanding of those ideas, build them into their own conceptual structures, and be able to illustrate them with examples and rational argument.

• Be comfortable and familiar with the standard vocabulary appropriate to the main ideas of science, mathematics, and technology, as used in this report. In many schools, science is taught solely as vocabulary, and that is largely what is tested. This approach is disastrous and is not what is called for here—which is a level of understanding of science that results in a useful vocabulary.

• Put correct interpretations on the terms "if . . . , then . . . ," "and," "every," "not," "correlates with," and "causes."

• Organize information into simple tables.

• Depict information and relationships by drawing freehand graphs to show trends (steady, accelerated, diminishing-return, and cyclic).

• Read values from pie charts and simple bar and line graphs, false-color maps, and two-way data tables, noting trends and extreme values and recognizing how the message in a graph is sensitive to the scale chosen.

• Check the correspondence between tabular, graphic, and verbal descriptions of data.

• Write and follow procedures in the form of step-by-step instructions, recipes, formulas, flow diagrams, and sketches.

• Comprehend and use basic geometrical relationships, including perpendicular, parallel, similar, congruent, tangent, rotation, and symmetry.

• Find and describe locations on maps, using rectangular and polar coordinates.

- Participate in group discussions on scientific topics by being able to restate or summarize what others have said, ask for clarification or elaboration, and take alternative perspectives.

Critical-Response Skills

In various forms, the mass media, teachers, and peers inundate students with assertions and arguments, some of them in the realm of science, mathematics, and technology. Education should prepare people to read or listen to such assertions critically, deciding what evidence to pay attention to and what to dismiss, and distinguishing careful arguments from shoddy ones. Furthermore, people should be able to apply those same critical skills to their own observations, arguments, and conclusions, thereby becoming less bound by their own prejudices and rationalizations.

Although most people cannot be expected to become experts in technical fields, everyone can learn to detect the symptoms of doubtful assertions and arguments. These have to do with the ways in which purported results are reported. Students should learn to notice and be put on their guard by the following signs of weak arguments:

- The premises of the argument are not made explicit.

- The conclusions do not follow logically from the evidence given (for example, the truth of "Most rich people vote Republican" does not prove the truth of the converse, "Most people who vote Republican are rich").

- The argument is based on analogy but the comparison is not apt.

- Fact and opinion are intermingled, opinions are presented as facts, or it is not clear which is which.

- Celebrity is used as authority ("Film star endorses new diet").

- Vague attributions are used in place of specific references (for example, such common attributions as "leading doctors say . . . ," "science has shown that . . . ," "compared to some other states . . . ," and "the scientific community recommends that . . .").

- No mention is made, in self-reported opinions or information, of measures taken to guard against deliberate or subconscious distortion.

- No mention is made, in evidence said to come from an experiment, of control groups very much like the experimental group.

- Graphs are used that—by chopping off part of the scale, using unusual scale units, or using no scale at all—distort the appearance of results.

- It is implied that all members of a group—such as "teenagers," "consumers," "immigrants," or "patients"—have nearly identical characteristics that do not overlap those of other groups.

- Average results are reported, but not the amount of variation around the average.

- A percentage or fraction is given, but not the total sample size (as in "9 out of 10 dentists recommend . . .").

- Absolute and proportional quantities are mixed (as in "3,400 more robberies in our city last year, whereas other cities had an increase of less than 1 percent").

- Results are reported with misleading preciseness (for example, representing 13 out of 19 students as 68.42 percent).

- Explanations or conclusions are represented as the only ones worth consideration, with no mention of other possibilities.

PART III:

BRIDGES TO THE FUTURE

Comet Halley, photographed from Las Campanas Observatory, Chile, March 1986.

INTRODUCTION

Project 2061 seeks to contribute to a major reform of science, mathematics, and technology education. It does so, though, in the full knowledge that this first report of the project will not be sufficient by itself to bring about nationwide reform. How, then, can the nation make significant headway toward realizing the goals expressed in this report? We think there are three conditions to be met: (1) improving the teaching of science, mathematics, and technology; (2) acquiring a realistic understanding of what it takes to achieve significant and lasting reform nationally; and (3) initiating immediate action on many fronts, on the basis of the recommendations of the National Council on Science and Technology Education. The views of Project 2061 on how to meet those conditions are presented below in three chapters:

• Chapter 13, "Effective Learning and Teaching," is based on the belief—strongly supported by the national council and the Phase I scientific panels—that how subjects are taught is as important as what subjects are taught. It presents, first, some well-established learning principles that should influence teaching generally, and then some principles of effective teaching that are consistent with and illuminate the national council's recommendations as to content.

• Chapter 14, "Educational Reform," looks at the question of educational reform in general. It discusses the national need for reform in science, mathematics, and technology education and outlines the premises that underlie the approach of Project 2061 to reform.

• Chapter 15, "Next Steps," outlines the purposes of Phase II of Project 2061 and then recommends how this report can be used by concerned individuals, institutions, and organizations to contribute to reform in science, mathematics, and technology education.

Jean-Baptiste-Siméon Chardin, *The Young Governess* (ca. 1739).

CHAPTER 13

EFFECTIVE LEARNING AND TEACHING

Although *Science for All Americans* emphasizes what students should learn, it also recognizes that how science is taught is equally important. In planning instruction, effective teachers draw on a growing body of research knowledge about the nature of learning and on craft knowledge about teaching that has stood the test of time. Typically, they consider the special characteristics of the material to be learned, the background of their students, and the conditions under which the teaching and learning are to take place.

This chapter presents—nonsystematically and with no claim of completeness—some principles of learning and teaching that characterize the approach of such teachers. Many of those principles apply to learning and teaching in general, but clearly some are especially important in science, mathematics, and technology education. For convenience, learning and teaching are presented here in separate sections, even though they are closely interrelated.

PRINCIPLES OF LEARNING

Learning Is Not Necessarily an Outcome of Teaching

Cognitive research is revealing that even with what is taken to be good instruction, many students, including academically talented ones, understand less than we think they do. With determination, students taking an examination are commonly able to identify what they have been told or what they have read; careful probing, however, often shows that their understanding is limited or distorted, if not altogether wrong. This finding suggests that parsimony is essential in setting out educational goals: Schools should pick the most important concepts and skills to emphasize so that they can concentrate on the quality of understanding rather than on the quantity of information presented.

What Students Learn Is Influenced by Their Existing Ideas

People have to construct their own meaning regardless of how clearly teachers or books tell them things. Mostly, a person does this by connecting new information and concepts to what he or she already believes. Concepts—the essential units of human thought—that do not have multiple links with how a student thinks about the world are not likely to be remembered or useful. Or, if they do remain in memory, they will be tucked away in a drawer labeled, say, "biology course, 1995," and will not be available to affect thoughts about any other aspect of the world. Concepts are learned best when they are encountered in a variety of contexts and expressed in a variety of ways, for that ensures that there are more opportunities for them to become imbedded in a student's knowledge system.

But effective learning often requires more than just making multiple connections of new ideas to old ones; it sometimes requires that people restructure their thinking radically. That is, to incorporate some new idea, learners must change the connections among the things they already know, or even discard some long-held beliefs about the world. The alternatives to the necessary restructuring are to distort the new information to fit their old ideas or to reject the new information entirely. Students come to school with their own ideas, some correct and some not, about almost every topic they are likely to encounter. If their intuition and misconceptions are ignored or dismissed out of hand, their original beliefs are likely to win out in the long run, even though they may give the test answers their teachers want. Mere contradiction is not sufficient; students must be encouraged to develop new views by seeing how such views help them make better sense of the world.

Progression in Learning Is Usually From the Concrete to the Abstract

Young people can learn most readily about things that are tangible and directly accessible to their senses—visual, auditory, tactile, and kinesthetic. With experience, they grow in their ability to understand abstract concepts, manipulate symbols, reason logically, and generalize. These skills develop slowly, however, and the dependence of most people on concrete examples of new ideas persists throughout life. Concrete experiences are most effective in learning when they occur in the context of some relevant conceptual structure. The difficulties many students have in grasping abstractions are often masked by their ability to remember and recite technical terms that they do not understand. As a result, teachers—from kindergarten through college—sometimes overestimate the ability of their students to handle abstractions, and they take the students' use of the right words as evidence of understanding.

People Learn to Do Well Only What They Practice Doing

If students are expected to apply ideas in novel situations, then they must practice applying them in novel situations. If they practice only calculating answers to predictable exercises or unrealistic "word problems," then that is all they are likely to learn. Similarly, students cannot learn to think critically, analyze information, communicate scientific ideas, make logical arguments, work as part of a team, and acquire other desirable skills unless they are permitted and encouraged to do those things over and over in many contexts.

Effective Learning by Students Requires Feedback

The mere repetition of tasks by students—whether manual or intellectual—is unlikely to lead to improved skills or keener insights. Learning often takes place best when students have opportunities to express ideas and get feedback from their peers. But for feedback to be most helpful to learners, it must consist of more than the provision of correct answers. Feedback ought to be analytical, to be suggestive, and to come at a time when students are interested in it. And then there must be time for students to reflect on the feedback they receive, to make adjustments and to try again—a requirement that is neglected, it is worth noting, by most examinations—especially finals.

Expectations Affect Performance

Students respond to their own expectations of what they can and cannot learn. If they believe they are able to learn something, whether solving equations or riding a bicycle, they usually make headway. But when they lack confidence, learning eludes them. Students grow in self-confidence as they experience success in learning, just as they lose confidence in the face of repeated failure. Thus, teachers need to provide students with challenging but attainable learning tasks and help them succeed.

What is more, students are quick to pick up the expectations of success or failure that others have for them. The positive and negative expectations shown by parents, counselors, principals, peers, and—more generally—by the media affect students' expectations and hence their learning behavior. When, for instance, a teacher signals his or her lack of confidence in the ability of students to understand certain subjects, the students may lose confidence in their ability and may perform more poorly than they otherwise might. If this apparent failure reinforces the teacher's original judgment, a disheartening spiral of decreasing confidence and performance can result.

TEACHING SCIENCE, MATHEMATICS, AND TECHNOLOGY

Teaching Should Be Consistent With the Nature of Scientific Inquiry

Science, mathematics, and technology are defined as much by what they do and how they do it as they are by the results they achieve. To understand them as ways of thinking and doing, as well as bodies of knowledge, requires that students have some experience with the kinds of thought and action that are typical of those fields. Teachers, therefore, should do the following:

Start With Questions About Nature. Sound teaching usually begins with questions and phenomena that are interesting and familiar to students, not with abstractions or phenomena outside their range of perception, understanding, or knowledge. Students need to get acquainted with the things around them—including devices, organisms, materials, shapes, and numbers—and to observe them, collect them, handle them, describe them, become puzzled by them, ask questions about them, argue about them, and then to try to find answers to their questions.

Engage Students Actively. Students need to have many and varied opportunities for collecting, sorting and cataloging; observing, note taking and sketching; interviewing, polling, and surveying; and using hand lenses, microscopes, thermometers, cameras, and other common instruments. They should dissect; measure, count, graph, and compute; explore the chemical properties of common substances; plant and cultivate; and systematically observe the social behavior of humans and other animals. Among these activities, none is more important than measurement, in that figuring out what to measure, what instruments to use, how to check the correctness of measurements, and how to configure and make sense out of the results are at the heart of much of science and engineering.

Concentrate on the Collection and Use of Evidence. Students should be given problems—at levels appropriate to their maturity—that require them to decide what evidence is relevant and to offer their own interpretations of what the evidence means. This puts a premium, just as science does, on careful observation and thoughtful analysis. Students need guidance, encouragement, and practice in collecting, sorting, and analyzing evidence, and in building arguments based on it. However, if such activities are not to be destructively boring, they must lead to some intellectually satisfying payoff that students care about.

Provide Historical Perspectives. During their school years, students should encounter many scientific ideas presented in historical context. It matters less which particular episodes teachers select (in addition to the few key episodes presented in Chapter 10) than that the selection represent the scope and diversity of the scientific enterprise. Students can develop a sense of how science really happens by learning something of the growth of scientific ideas, of the twists and turns on the way to our current understanding of such ideas, of the roles played by different investigators and commentators, and of the interplay between evidence and theory over time.

History is important for the effective teaching. of science, mathematics, and technology also because it can lead to social perspectives—the influence of society on the development of science and technology, and the impact of science and technology on society. It is important, for example, for students to become aware that women and minorities have made significant contributions in spite of the barriers put in their way by society; that the roots of science, mathematics, and technology go back to the early Egyptian, Greek, Arabic, and Chinese cultures; and that scientists bring to their work the values and prejudices of the cultures in which they live.

Insist on Clear Expression. Effective oral and written communication is so important in every facet of life that teachers of every subject and at every level should place a high priority on it for all students. In addition, science teachers should emphasize clear expression, because the role of evidence and the unambiguous replication of evidence cannot be understood without some struggle to express one's own procedures, findings, and ideas rigorously, and to decode the accounts of others.

Use a Team Approach. The collaborative nature of scientific and technological work should be strongly reinforced by frequent group activity in the classroom. Scientists and engineers work mostly in groups and less often as isolated investigators. Similarly, students should gain experience sharing responsibility for learning with each other. In the process of coming to common understandings, students in a group must frequently inform each other about procedures and meanings, argue over findings, and assess how the task is progressing. In the context of team responsibility, feedback and communication become more realistic and of a character very different from the usual individualistic textbook-homework-recitation approach.

Do Not Separate Knowing From Finding Out. In science, conclusions and the methods that lead to them are tightly coupled. The nature of inquiry depends on what is being investigated, and what is learned depends on the methods used. Science teaching that attempts solely to impart to students the accumulated knowledge of a field leads to very little understanding and certainly not to the development of intellectual independence and facility. But then, to teach scientific reasoning as a set of procedures separate from any particular substance—"the scientific method," for instance—is equally futile. Science teachers should help students to acquire both scientific knowledge of the world and scientific habits of mind at the same time.

Deemphasize the Memorization of Technical Vocabulary. Understanding rather than vocabulary should be the main purpose of science teaching. However, unambiguous terminology is also important in scientific communication and—ultimately—for understanding. Some technical terms are therefore helpful for everyone, but the number of essential ones is relatively small. If teachers introduce technical terms only as needed to clarify thinking and promote effective communication, then students will gradually build a functional vocabulary that will survive beyond the next test. For teachers to concentrate on vocabulary, however, is to detract from science as a process, to put learning for understanding in jeopardy, and to risk being misled about what students have learned.

Science Teaching Should Reflect Scientific Values

Science is more than a body of knowledge and a way of accumulating and validating that knowledge. It is also a social activity that incorporates certain human values. Holding curiosity, creativity, imagination, and beauty in high esteem is certainly not confined to science, mathematics, and engineering—any more than skepticism and a distaste for dogmatism are. However, they are all highly characteristic of the scientific endeavor. In learning science, students should encounter such values as part of their experience, not as empty claims. This suggests that teachers should strive to do the following:

Welcome Curiosity. Science, mathematics, and technology do not create curiosity. They accept it, foster it, incorporate it, reward it, and discipline it—and so does good science teaching. Thus, science teachers should encourage students to raise questions about the material being studied, help them learn to frame their questions clearly enough to begin to search for answers, suggest to them productive ways for finding answers, and reward those who raise and then pursue unusual but relevant questions. In the science classroom, wondering should be as highly valued as knowing.

Reward Creativity. Scientists, mathematicians, and engineers prize the creative use of imagination. The science classroom ought to be a place where creativity and invention—as qualities distinct from academic excellence—are recognized and encouraged. Indeed, teachers can express their own creativity by inventing activities in which students' creativity and imagination will pay off.

Encourage a Spirit of Healthy Questioning. Science, mathematics, and engineering prosper because of the institutionalized skepticism of their practitioners. Their central tenet is that one's evidence, logic, and claims will be questioned, and one's experiments will be subjected to replication. In science classrooms, it should be the normal practice for teachers to raise such questions as: How do we know? What is the evidence? What is the argument that interprets the evidence? Are there alternative explanations or other ways of solving the problem that could be better? The aim should be to get students into the habit of posing such questions and framing answers.

Avoid Dogmatism. Students should experience science as a process for extending understanding, not as unalterable truth. This means that teachers must take care not to convey the impression that they themselves or the textbooks are absolute authorities whose conclusions are always correct. By dealing with the credibility of scientific claims, the overturn of accepted scientific beliefs, and what to make out of disagreements among scientists, science teachers can help students to balance the necessity for accepting a great deal of science on faith against the importance of keeping an open mind.

Promote Aesthetic Responses. Many people regard science as cold and uninteresting. However, a scientific understanding of, say, the formation of stars, the blue of the sky, or the construction of the human heart need not displace the romantic and spiritual meanings of such phenomena. Moreover, scientific knowledge makes additional aesthetic responses possible—such as to the diffracted pattern of street lights seen through a curtain, the pulse of life in a microscopic organism, the cantilevered sweep of a bridge, the efficiency of combustion in living cells, the history in a rock or a tree, an elegant mathematical proof. Teachers of science, mathematics, and technology should establish a learning environment in which students are able to broaden and deepen their response to the beauty of ideas, methods, tools, structures, objects, and living organisms.

Science Teaching Should Aim to Counteract Learning Anxieties

Teachers should recognize that for many students, the learning of mathematics and science involves feelings of severe anxiety and fear of failure. No doubt this results partly from what is taught and the way it is taught, and partly from attitudes picked up incidentally very early in schooling from parents and teachers who are themselves ill at ease with science and mathematics. Far from dismissing math and science anxiety as groundless, though, teachers should assure students that they understand the problem and will work with them to overcome it. Teachers can take such measures as the following:

Build on Success. Teachers should make sure that students have some sense of success in learning science and mathematics, and they should deemphasize getting all the right answers as being the main criterion of success. After all, science itself, as Alfred North Whitehead said, is never quite right. Understanding anything is never absolute, and it takes many forms. Accordingly, teachers should strive to make all students—particularly the less-confident ones—aware of their progress and should encourage them to continue studying.

Provide Abundant Experience in Using Tools. Many students are fearful of using laboratory instruments and other tools. This fear may result primarily from the lack of opportunity many of them have to become familiar with tools in safe circumstances. Girls in particular suffer from the mistaken notion that boys are naturally more adept at using tools. Starting in the earliest grades, all students should gradually gain familiarity with tools and the proper use of tools. By the time they finish school, all students should have had supervised experience with common hand tools, soldering irons, electrical meters, drafting tools, optical and sound equipment, calculators, and computers.

Support the Roles of Girls and Minorities in Science. Because the scientific and engineering professions have been predominantly male and white, female and minority students could easily get the impression that these fields are beyond them or are otherwise unsuited to them. This debilitating perception—all too often reinforced by the environment outside the school—will persist unless teachers actively work to turn it around. Teachers should select learning materials that illustrate the contributions of women and minorities, bring in role models, and make it clear to female and minority students that they are expected to study the same subjects at the same level as everyone else and to perform as well.

Emphasize Group Learning. A group approach has motivational value apart from the need to use team learning (as noted earlier) to promote an understanding of how science and engineering work. Overemphasis on competition among students for high grades distorts what ought to be the prime motive for studying science: to find things out. Competition among students in the science classroom may also result in many of them developing a dislike of science and losing their confidence in their ability to learn science. Group approaches, the norm in science, have many advantages in education; for instance, they help youngsters see that everyone can contribute to the attainment of common goals and that progress does not depend on everyone's having the same abilities.

Science Teaching Should Extend Beyond the School

Children learn from their parents, siblings, other relatives, peers, and adult authority figures, as well as from teachers. They learn from movies, television, radio, records, trade books and magazines, and home computers, and from going to museums and zoos, parties, club meetings, rock concerts, and sports events, as well as from schoolbooks and the school environment in general. Science teachers should exploit the rich resources of the larger community and involve parents and other concerned adults in useful ways. It is also important for teachers to recognize that some of what their students learn informally is wrong, incomplete, poorly understood, or misunderstood, but that formal education can help students to restructure that knowledge and acquire new knowledge.

Teaching Should Take Its Time

In learning science, students need time for exploring, for making observations, for taking wrong turns, for testing ideas, for doing things over again; time for building things, calibrating instruments, collecting things, constructing physical and mathematical models for testing ideas; time for learning whatever mathematics, technology, and science they may need to deal with the questions at hand; time for asking around, reading, and arguing; time for wrestling with unfamiliar and counterintuitive ideas and for coming to see the advantage in thinking in a different way. Moreover, any topic in science, mathematics, or technology that is taught only in a single lesson or unit is unlikely to leave a trace by the end of schooling. To take hold and mature, concepts must not just be presented to students from time to time but must be offered to them periodically in different contexts and at increasing levels of sophistication.

Arshile Gorky, *Organization* (1933/1936).

CHAPTER 14

REFORMING EDUCATION

Project 2061 is concerned more with lasting reform of education than with the immediate improvement of the schools—although such improvement is certainly needed, possible, and under way in many parts of the United States. But, as the nation discovered after *Sputnik* more than three decades ago, enduring educational reform is not easily achieved.

The possibility of successfully restructuring science education in its entirety depends on the presence of a public demand for reform in science education and on what we as a nation think it takes to achieve reform. This chapter begins by showing that there is in fact a consensus on the need for reform in science, mathematics, and technology education, and then presents the premises that underlie the approach of Project 2061 to reform.

THE NEED FOR REFORM

The necessity for strengthening science education in the United States has been widely acknowledged in the numerous education studies conducted in the 1980s (a representative selection of reports is listed in Appendix B). Although the most powerful argument for improving the science education of all students may be its role in liberating the human intellect, much of the public discussion has centered on more concrete, utilitarian, and immediate justifications.

Most of the education reports of the 1980s have been motivated by the confluence of two different growing public concerns. One concern is America's seeming economic decline. Our domestic affluence and international power—both based substantially on our scientific and technological preeminence—have been weakening in relation to those of other countries, especially Japan. The other concern consists of certain trends in U.S. public education: low test scores, students' avoidance of science and mathematics, a demoralized and weakening teaching staff in many schools, low learning expectations compared to other technologically advanced nations, and being ranked near the bottom in international studies of students' knowledge of science and mathematics. All of the reports and the mass media coverage of the reports have highlighted these educational deficiencies, and the nation has finally become aware that indeed there is a crisis in American education.

Even while being deplored for themselves, the educational failures in the United States have come to be seen collectively as a major contributor to the economic failures. This view, whether entirely justified or not, has been implicit in most of the reports and explicit in others. Although each of the various reports has addressed the issues from a somewhat different perspective, all have been energized by the same set of disturbing economic and educational trends.

Given this background, it is understandable that the reports emphasize, in one way or another, the need to improve the science and technology education of all students, as well as the need for various educational reforms of a more general nature. Taken together, the reports serve to underscore that in our postindustrial society, there is a strong connection between how well a nation can perform and the existence of high-quality, widely distributed education. There is now a clear national consensus in the United States that all elementary and secondary school children need to become better educated in science, mathematics, and technology.

REFORM PREMISES

Reform Necessarily Takes a Long Time

Quick fixes always fail in education, and for readily understandable reasons. Perhaps the most obvious of these is simply the size of the enterprise. Education in the United States is an enormous business, employing more than 3 million people, expending nearly $200 billion a year, and holding collective capital assets in excess of $1 trillion. It is quixotic to believe that elementary and secondary education in America—serving nearly 50 million students located in more than 80,000 schools and 50 states—could easily or quickly be changed. Even with great ideas, the best of intentions, an investment of resources on a scale appropriate to the job, and lots of hard work, any sweeping change in the educational system nationally is bound to take a decade or longer.

It is more than simply a problem of scale, however. Unlike the situation in most other countries, the system of education in the United States is decentralized politically and economically. Decisions on educational policy and the use of resources for education are made by literally thousands of different entities, including 16,000 separate school districts, 3,300 colleges and universities, 50 states, several agencies of the federal government, and the courts at every level. This state of affairs may have its advantages, but a capacity for rapid change is not one of them. It takes time, first of all, for a strong consensus to build among educators and the public that radical change is needed. Then more time is needed to come to some national meeting of the minds on what the main ingredients of reform should be. Still more time is needed for action plans to be drawn up, ideas tested, and action initiated in tens of thousands of different institutions.

Ultimately, reform is more about people than it is about policies, institutions, and processes. And most people—not only educators— tend to change slowly when it comes to attitudes, beliefs, and ways of doing things. Teachers and administrators bring to their jobs the full range of human views about the purposes of education, the nature of young people, and the best ways to foster learning. Their views have been derived from and reinforced by years of experience—as students, teachers, and, often, parents. Sensible professionals do not replace their strongly held views and behavior patterns in response to fiat or the latest vogue; instead, they respond to developing sentiment among respected colleagues, to incentives that reward serious efforts to explore new possibilities, and to the positive feedback that may come from trying out new ideas from time to time—all of which can take years.

Professions may change mostly in response to turnover. Young physicians and engineers, for instance, carry new knowledge, techniques, and attitudes into those professions. Successive generations of teachers and school administrators can serve in the same way, but only if they come bearing different attitudes, knowledge, and skills than the ones they replace. Reforming teachers' education, therefore, is the sine qua non of school reform, but it will necessarily be slow to make its impact felt.

Collaboration Is Essential

Monolithic approaches to educational reform are not the American way, and with good reason: No group or sector is in sole possession of wisdom, inventiveness, resources, and authority, and few educational problems of consequence have only one possible solution. But diversity of effort can lead to little impact on a national scale if those who are striving to change things are all heading in different directions without regard for each other. Lockstep in education is neither possible nor desirable, but a commitment to collaboration is. Operationally, such a commitment means sharing ideas and information with others who are addressing the same or related problems. In the context of the reform of science education, this observation applies to the scientific community itself to the degree it wishes to make significant contributions to the process of reform in education.

Project 2061 constitutes, of course, only one of many efforts to chart new directions in science, mathematics, and technology education and to bring about significant improvements in the current system. Here and there across the nation, individual teachers and schools are striving, often against heavy odds, to change things, and in some school districts and states, vigorous reform is now the order of the day. Moreover, on a national scale, there are projects—many of them funded by foundations and government agencies and centered in professional associations, universities, and independent organizations—that are focusing on various aspects of reform. There is a need for these various reform efforts to link up to bring coherence to the movement.

Teachers Are Central

Although creative ideas for reforming education come from many sources, only teachers can provide the insights that emerge from intensive, direct experience in the classroom itself. They bring to the task of reform a knowledge of students, craft, and school culture that others cannot. Moreover, reform cannot be imposed on teachers from the top down or the outside in. If teachers are not convinced of the merit of proposed changes, they are unlikely to implement them energetically. If they do not understand fully what is called for or have not been sufficiently well prepared to introduce new content and ways of teaching, reform measures will founder. In either case, the more teachers share in shaping reform measures and the more help they are given in implementing agreed-upon changes, the greater the probability that they will be able to make those improvements stick.

Although teachers are central to reform, they cannot be held solely responsible for achieving it. They need allies. Teachers alone cannot change the textbooks, install more sensible testing policies than are now in place, create administrative support systems, get the public to understand where reform is headed and why it takes time to get there, and raise the funds needed to pay for reform. Thus, school administrators and education policymakers need to support teachers. Teachers also need academic colleagues—scholars who are experts on relevant subject matter, child development, learning, and the educational potential of modern technologies. And they need the help and support of community leaders, business and labor leaders, and parents—for in the final analysis, educational reform is a shared responsibility. It is time for teachers to take more responsibility for the reform of education, but that in no way reduces the responsibility of others to do their part too.

Comprehensive Approaches Are Needed

Piecemeal reform measures beget piecemeal effects, if any. At the school district level, reform efforts should be inclusive: all grades, all subject domains, all streams. It is less demanding to concentrate on, say, improving third-grade reading, junior high school social studies, and biology for vocational students. But such unrelated changes are not likely to add up to curricula that are any more integrated, coherent, and effective than the fragmented, overburdened ones that now exist. Without a more sweeping approach, change will be constrained by having to fit within the boundaries of class periods, school subjects, sequences, and tracks that themselves may be a large part of the problem.

Nationwide, reform needs to be comprehensive in the sense of addressing all aspects of the system. Reform in science education depends on changing existing curricula from kindergarten through high school. But to make new curricula work, changes must also occur in the preparation of teachers, the content of textbooks and other learning materials, the use of technologies, the nature of testing, and the organization of schools. Furthermore, the changes need to be compatible, lest they cancel each other out.

Comprehensive reform does not imply going off in all directions at once. Rather, it demands that some steps occur before others, that some problems take precedence, and that resources be deployed strategically. Careful systemwide planning should precede action, and no aspect of planning is more crucial than setting priorities. Failure to set priorities can result in only a little change; setting the wrong priorities may leave the students worse off than before reform was undertaken.

Reform Must Focus on the Science Learning Needs of All Children

When demographic realities, national needs, and democratic values are taken into account, it becomes clear that the nation can no longer ignore the science education of any students. Race, language, sex, or economic circumstances must no longer be permitted to be factors in determining who does and who does not receive a good education in science, mathematics, and technology. To neglect the science education of any (as has happened too often to girls and minority students) is to deprive them of a basic education, handicap them for life, and deprive the nation of talented workers and informed citizens—a loss the nation can ill afford.

To reach all students means reforming the education of every strand of the student body—vocational, general, and college preparatory. For students who expect to go right to work after high school, a narrow focus on trade skills will no longer do; they need to acquire a strong base of scientific knowledge and of reasoning, communication, and learning skills. All college-bound students, quite apart from what they believe their majors will eventually turn out to be, need to enter college with an understanding of science, mathematics, and technology that they can build on and that will make it possible for them to elect a technical field. And undecided students need the knowledge, skills, and attitudes to enable them to move in any direction. The recommendations in this report, therefore, apply equally to all students.

Meeting the science learning needs of all children requires that society as a whole recognize that learning is, in a sense, the chief occupation of childhood. Play is important for its own sake and because it often leads to learning, and work for money can be instructive for children, but neither play nor employment can substitute for systematic study. Parents and citizens in general, therefore, must understand that a substantial portion of the energies of childhood have to be devoted to the task of learning.

Positive Conditions for Reform Must Be Established

Reform requires creating conditions for change. There is no sense in exhorting educators to change what they are doing and then ignoring the obstacles in their path. Not surprisingly, a major barrier to reform is the same barrier that gets in the way of good education in general: the working circumstances of teachers and administrators.

In all too many schools, physical, administrative, and psychological circumstances militate against undertaking major curricular reform efforts. Typically, teachers lack time to think, study, organize materials, confer with colleagues, counsel individual students, and attend professional meetings. What is more, they do not have private offices, computers for word processing and recordkeeping, laboratory assistants, access to expert consultants, or the other kinds of support that professionals in other fields expect. And principals are scarcely better off. The press of such demanding matters as public relations, personnel management, budgets, student attendance, and safety leave principals with little time, energy, or inclination to engage in program matters at all—let alone in major reform activities.

At the same time as barriers to reform are being removed, positive conditions for change must be established. They need to emphasize creating an environment for teachers and administrators that encourages experimentation, focus on long-term gains rather than on such immediate goals as raising test scores, and recognize and reward innovation.

The need for positive conditions for reform goes well beyond the schools. What schools can accomplish for many children is very limited as long as a quarter of the students are raised in poverty, drug use and violence go unabated, racism persists, and commercial television remains vapid or worse while educational television stays chronically undernourished. It is an admirable notion that better education is necessary for and can lead to a better America. But only if some of today's worst social problems are ameliorated will the schools be able to take the sweeping reform steps that will enable them to have extensive positive effects on society. The reforming of education and the reforming of society need to go hand in hand.

To help ensure that reform does happen, continuing community support for education is essential. Such support is not easy to sustain in the face of changing demographics and changing social priorities. Therefore, informed and determined political leadership at every level and in every sector—government, business, labor, and education—is crucial for achieving reform. Without such leadership, community support for educational reform will fade away long before lasting results can be achieved.

Maurits Cornelis Escher, *Liberation* (1955).

CHAPTER 15
NEXT STEPS

This Project 2061 report has little to say about what ails the educational system, points no finger of blame, prescribes no specific remedies. Rather, it attempts to contribute substantially to educational reform by serving as a starting point for two sets of critical, reform-oriented actions.

One set is based on use of the report as the first step in a multistage, long-term developmental process. *Science for All Americans,* along with the Phase I panel reports, is being used as the conceptual basis for activities in Phase II of Project 2061. Those activities will result in recommendations for change in all parts of the educational system—changes that will be implemented in Phase III of the project.

The other set of actions is based on the fact that the report provides a new and unusually substantive opportunity for everyone who has a stake in educational reform to reappraise the progress made so far, redirect their efforts as needed, and recommit themselves to fundamental reform goals.

This final chapter of *Science for All Americans* starts with a brief outline of the next steps toward reform being taken by Project 2061 in Phase II. It then explores some of the ways in which the report can be put to work by educators, policymakers, and the interested public.

PHASE II OF PROJECT 2061

The second phase of Project 2061 is a direct extension of the first phase. Its main purposes are as follows:

• To produce a diverse array of curriculum models for kindergarten through high school (K-12) based on the recommendations of this report.

• To create a set of blueprints for reforming the other components of education that complement curriculum reform.

• To increase the pool of educators and scientists able to serve as experts in school curriculum reform.

• To foster public awareness of the need for reform in science, mathematics, and technology education, and to promote reform efforts among teachers, administrators, and education policymakers.

Curriculum Models

The main creative activity of Phase II of Project 2061 is to develop, in five school districts across the nation, alternative K-12 curriculum models for education in science, mathematics, and technology. The development team in each district will include teachers from all grades, from the physical, biological, and social sciences, and from mathematics and technology. The new curriculum models will all be aimed at achieving the recommendations of this report, but they will differ from one another in other ways. They are expected to vary in emphasis, style, and degree to which they diverge from current models.

As the models are being created, a standard format will be developed for describing K-12 curricula in science, mathematics, and technology. If successful, this will make it possible, as it is not now, to characterize and compare the curricula of different school districts by highlighting their key features.

Blueprints for Action

New curriculum models by themselves can no more bring about actual reform than can a consensus on learning goals. Both are necessary, but not sufficient. Consequently, in Phase II, the project members will work with others to create blueprints for achieving national reform in science, mathematics, and technology education. In a series of reports, they will offer recommendations concerning the education of teachers, policies and instruments to be used in testing, educational materials and technologies, the structure of schooling and the organization of instruction, education policy, educational research, and implementation strategies.

Curriculum Reform Experts

It takes people to change systems. Actually changing curricula in science, mathematics, and technology to reflect the goals of this report will not happen automatically—no matter how appealing the new Phase II curriculum models may turn out to be. Successful implementation, in Phase III, will depend upon the existence of a cadre of committed, knowledgeable, and experienced leaders. Accordingly, one of the goals of Phase II is to create a pool of educators and scientists who are broadly conversant with the contents of the national council's recommendations and are also skilled in translating such material into actual curricula.

Promoting Reform

During Phase II, various steps will be taken to foster discussion of the need for reform in science, mathematics, and technology education and of what has to be done to achieve it. These steps will include the widespread dissemination of *Science for All Americans;* articles in professional and popular journals; workshops and seminars at professional meetings; the dissemination of the blueprint-for-action reports to educators, scientists, and the media; and the preparation of a series of papers directed to the attention of particular audiences, such as primary grade teachers, middle school principals, high school social studies teachers, or school board members.

AN AGENDA FOR ACTION

The following comments are meant to provoke action and debate. The greater the number of individuals, institutions, and organizations that become engaged in discussing what they can do to contribute to the reform of science education—and then follow up their plans with action—the sooner the nation will begin to make progress.

Public Support

Truly fundamental reform in science, mathematics, and technology education is possible only if there is widespread public support for it. This report can be used to help secure such support and to cast it in the context of desired goals rather than particular mechanisms. To that end, Project 2061 recommends that

• The President of the United States use this report, along with others, as a basis for speaking forcefully to the American people on the need for scientific literacy; establish scientific literacy as a national goal; and periodically reinforce the priority of the goal.

• The U.S. secretary of education publicly support and elaborate the theme of scientific literacy; encourage the development of techniques for sampling and measuring meaningful learning that will allow monitoring progress toward scientific literacy; and announce that progress toward scientific literacy will become part of the nation's annual "Report Card."

• Congress pass a joint resolution indicating to the public its own concern over the weak state of science, mathematics, and technology education in this country; and conduct hearings to identify what steps it might take to help the nation reach the goals of this report.

• The governors of all of the states issue public statements establishing scientific literacy as a priority and signaling their intention to press for needed reforms; and use the National Governors' Association and the Education Commission of the States to place this report on their agendas for debate.

• Business and labor leaders of the nation speak out individually and through their organizations on the urgent need for all Americans to have the knowledge and skills set out in this report, and pledge their support of efforts to reform science education.

• The news media bring the recommendations of this report to the attention of the public by having leading scientists, educators, business and labor executives, military officers, elected officials, and social commentators discuss and debate them on radio and television and in newspapers and popular magazines.

Educational Leadership

Reform also depends on the readiness of teachers, school administrators, and education policymakers to support it and to provide leadership. They will do so only if they become convinced that scientific literacy should be a basic requirement for all children and that the goals defining scientific literacy make good educational sense. This report is well suited to serve as the vehicle for getting educators behind a national effort to reform science, mathematics, and technology education. Accordingly, Project 2061 recommends that

• The secretary of education encourage all state and local education agencies to assign a high priority to the universal attainment of scientific literacy; require the appropriate assistant secretaries in the Department of Education to find ways in which their programs can contribute to that goal; and initiate mechanisms to help states and inner-city districts develop and carry out plans to bring minorities and other educationally disadvantaged youth up to the standards recommended in this report.

- Each state board of education set up a blue-ribbon panel to examine *Science for All Americans* and to report on its educational implications to the chief state school officer, the state legislature, local school boards, and to the state's school superintendents, principals, and teachers.

- All national education associations—including those of teachers, school administrators, school boards and parents—report to their members on the recommendations of this report, promote debate on them, and establish mechanisms for fostering the recommendations that they support.

- The National Science Teachers Association, the National Council of Teachers of Mathematics, the National Council for the Social Studies, the International Technology Education Association, and allied teaching societies take the lead in fostering the goals of *Science for All Americans* among teachers of all subjects and levels, administrators, and education policymakers.

Collaboration

Educational reform must be collaborative to succeed. In the case of science, mathematics, and technology education, the scientific community must enter into partnership with the education community. Although several hundred scientists, engineers, and mathematicians participated in framing the recommendations of this report, it will take the involvement of many more as the reform movement gains momentum. To that end, Project 2061 recommends that

- The heads of the National Science Foundation, the National Institutes of Health, the National Bureau of Standards, the National Aeronautics and Space Administration, the Department of Energy, the Department of Agriculture, and other science-related federal agencies and departments impress on their constituencies the need to help educators improve education in science, mathematics, and technology; and require their staffs to develop appropriate ways in which their agencies can contribute to that effort.

- All national scientific, engineering, mathematical, and medical societies and state academies of science use this report to stimulate discussion among their members on what constitutes scientific literacy; ask their members to work with educators toward shared goals; and use this report in formulating plans for helping educators.

- The Triangle Coalition, the Coalition for Education in the Sciences, the state alliances for science, and other groups—which already bring together leaders from the scientific, educational, and business communities—determine ways in which this report can be used to further the participation of scientists in their own reform efforts.

- The National Science Teachers Association, the National Council for the Social Studies, the National Council of Teachers of Mathematics, and the International Technology Education Association form a joint commission to consider what collaborative actions teachers of those subjects might take to support the recommendations of this report that cut across fields.

Qualified Teachers

The scientific literacy goals of this report can be reached only if students in elementary and secondary school have teachers who are fully qualified to teach. Sad to say, that is all too often not now the case—and all the more regrettable in light of the breadth and depth of understanding of science, mathematics, and technology called for here. Thus, Project 2061 recommends that

• Teachers stand solidly behind efforts—such as those of the National Board for Professional Teaching Standards, the NSTA National Teacher Certification Program, and the Holmes Group—to raise standards for teaching in every field; and call upon those groups to use the recommendations of this report in establishing standards for science and mathematics teachers.

• The National Science Foundation, the National Science Teachers Association, and the National Council of Teachers of Mathematics review the criteria for selecting recipients of the Presidential Awards for Excellence in Science and Mathematics Teaching in the light of the recommendations of this report.

• Science, mathematics, and technology teaching associations stand behind efforts such as the Stanford Teacher Assessment Project to develop reliable ways for judging the capability of individuals to effectively teach the content outlined in this report.

• The National Science Foundation and the Department of Education seek budget support to enable them to accelerate the process of bringing the quality of the nation's teachers of science, mathematics, and technology up to the level of understanding set out in this report.

• The presidents of all colleges and universities establish scientific literacy as an institution-wide priority; and direct their institutions to reshape undergraduate requirements as necessary to ensure that all graduates (from whom, after all, tomorrow's teachers will be drawn) leave with an understanding of science, mathematics, and technology that surpasses what this report recommends for all high school graduates.

• College departments of science and mathematics use this report as a guide in designing courses for future elementary school teachers and high school science teachers that go beyond, but are in the spirit of, the recommendations of this report; and create and seek funding for the conduct of in-service workshops and institutes tailored to the needs of teachers who wish to attain the standard of excellence implicit in the recommendations presented in this report.

• Education faculties review the content and pedagogical standards for the preparation of elementary and secondary teachers of science in light of this report; and work with their colleagues in the other departments to institute changes in the way in which future teachers are prepared.

Instructional Materials

For teachers to be able to bring all students to the level of understanding and skill proposed in this report, they will need a new generation of books and other instructional tools. As in other complex undertakings, reaching demanding goals in education requires having access to appropriate technologies. Textbooks and other teaching materials in current use are—to put it starkly—simply not up to the job; and the potential of computers and other modern technologies has yet to be realized. Because this report is intended to add new dimensions to what teaching is supposed to achieve, and therefore to what kinds of materials will be needed, Project 2061 recommends that

• Textbook publishers convene a national meeting of senior science, mathematics, and technology editors to explore the substance of this report and discuss its implications for the future of the industry; and individually consider the report's recommendations and monitor the developing Phase II curriculum models as they plan future editions of existing books and decide which new ones they should begin to develop.

• Companies engaged in the production and sale of audiovisual educational materials for school, home, and library use this report as a guide in developing new products.

• The National Science Foundation again take the lead in supporting research and development on the use of computers and advanced interactive systems for teaching and learning, and significantly increase its budget for that purpose.

• The National Science Teachers Association, the National Council of Teachers of Mathematics, the National Council for Social Studies, and the International Technology Education Association cooperate in discussing with the developers of computer software what kinds of software teachers will need to teach the ideas and skills recommended in this report.

• The producers of educational achievement tests review this report with an eye to how the content and style of their instruments would need to be modified so that they could become incentives for purposeful learning of the kind presented here; and invest more heavily than in the past in developing new kinds of tests to provide practical alternatives to tests that reward only the memorization of bits of information.

Research

Finally, it ought to be understood that too little is known about how different kinds of children learn and about how to organize instruction for optimal effectiveness for anyone to be able to prescribe how best to achieve the goals presented in this report. For that reason, it is recommended that

• Both the Department of Education and the National Science Foundation dramatically increase their support of research related to the learning and teaching of science, mathematics, and technology; increase the proportion of research funding devoted to the support of research teams composed of outstanding natural and social scientists, mathematicians, engineers, cognitive and developmental psychologists, and educators to enable them to pursue productive lines of investigation over an extended period of time; and base their research agendas in part on the vision of scientific literacy presented in this report.

- The Department of Education make it possible for a few major cities with large populations of disadvantaged youth to redesign and reorganize their school systems radically, completely, and quickly as a large-scale, closely monitored national experiment to determine what is possible when the nation treats school reform with the same intensity, urgency, and application of resources that it applies to other national problems of great consequence.

THE FUTURE

What will this all add up to? Where will the nation be in, say, 1992, as Phase II comes to an end? Certainly none of our major educational problems will have been completely solved. Most students will still not be emerging from our schools well educated in science, mathematics, and technology. The nation's curricula will not be very different from what they are now. Nor will the textbooks, tests, and the rest of the components of education have been radically changed. And yet the need for scientifically literate citizens will surely be greater than ever by then.

But progress will have been made if, by 1992,

- The nation is still paying attention to educational reform in science, mathematics, and technology.

- Our national leaders are speaking out regularly and forcefully about the need for everyone to continue to pull together in the pursuit of scientific literacy.

- We have made up our minds about what we want to achieve in science, mathematics, and technology education—an end to which this report is intended to contribute.

- Educators and education policymakers have begun to develop a strong consensus on what it will take to restructure the school system so that all students—including especially those it has failed in the past—will emerge well educated in science, mathematics, and technology.

- We find that a large number of educators and scientists are actually collaborating in reform activities in school systems across the land and that their numbers are rapidly increasing.

- Scientists, educators, parents, and citizens have paid enough attention to this report to have identified its shortcomings and have taken the trouble to advise Project 2061 and other users on how to overcome them as we work together to improve the science, mathematics, and technology education of all Americans.

There are no valid reasons—intellectual, social, or economic— why the United States cannot transform its schools to make scientific literacy possible for all students. What is required is national commitment, determination, and a willingness to work together toward common goals. We trust that *Science for All Americans* clarifies those goals.

George W. Wetherill, the director of the Carnegie Institution of Washington's Department of Terrestrial Magnetism, wrote the following lines when he was in La Serena, Chile, in April 1986 to observe Comet Halley:

Among the eucalyptus trees,
Green leaves dancing in the autumn wind,
The cold pale watcher of mankind
Treads his ancient trail again.

Pass swiftly by the angry bull,
The starry fish and water jar,
Defy the Sun's consuming flame,
The archer's bow,
The scorpion's sting,
The centaur's wrath,
The deadly coil of the hydra—
But then be gone.
Ask not for Harold of Hastings,
You know he is not here;
Nor Attila, vanquished at Chalons,
Edmund, master of Isaac's rules.
Nor Giotto, and the Zealots of Jerusalem.

You must have seen
The ships that rose to greet you.
Next time there will be more.
They'll even mount your haggard head
And ride you into Neptune's night!
Yes, we still are bold.
Though once more we now learn
The message that you bear,
Resonate to your grim tattoo,
The gravest rhythm of our race,
Yet wait with hope your sure return.

APPENDIXES

APPENDIX A
PROJECT 2061 PHASE I PARTICIPANTS

Science for All Americans is the result of the efforts of more than 300 scientists and educators. Over a period of three years, these individuals contributed to the project in different ways and at different times, but always with a high level of thoughtfulness, professional integrity, and commitment to a future in which all children can achieve scientific literacy. This statement, however, should not be taken to imply that the individual participants endorse all of the recommendations in *Science for All Americans* or that they are in any way responsible for any of the report's errors or other shortcomings.

The participants are listed below in the following order:

- National Council on Science and Technology Education
- Members of the Phase I scientific panels (and panel staff members)
- Project 2061 staff (AAAS)
- Advisers to the Project 2061 staff
- Consultants to the Phase I panels
- Reviewers of *Science for All Americans*
- Reviewers of the Phase I panel reports

NATIONAL COUNCIL ON SCIENCE AND TECHNOLOGY EDUCATION

The members of the council are listed at the front of this report.

PHASE I PANELS

Biological and Health Sciences

Helen M. Ranney (Panel Chair) Professor of Medicine, University of California, San Diego; Distinguished Physician, VA Medical Center (La Jolla, California)

Mary Clark Professor of Biology, San Diego State University

Maxwell Cowan Vice President and Chief Scientific Officer, Howard Hughes Medical Institute

Tony Hunter Professor, Molecular Biology and Virology Laboratory, The Salk Institute

John Moore Professor of Biology, University of California, Riverside

George Somero Professor of Marine Biology, Scripps Institution of Oceanography

Nicholas Spitzer Professor of Biology, University of California, San Diego

Herbert Stern Professor of Biology, University of California, San Diego

Audrey Terras Professor of Mathematics, University of California, San Diego

Panel Staff: Jan Yaffe, Psychotherapist (San Diego)

Mathematics

David Blackwell (Panel Cochair) Professor of Statistics and Professor of Mathematics, University of California, Berkeley

Leon Henkin (Panel Cochair) Professor of Mathematics, University of California, Berkeley

Lenore Blum Professor of Mathematics, Mills College; Research Scientist, International Computer Science Institute

Paul Garabedian (Physical and Information Sciences and Engineering Panel Liaison Member) Professor of Mathematics, Courant Institute of Mathematical Sciences, New York University

Paul Halmos Professor of Mathematics, University of Santa Clara

Harvey Keynes (Technology Panel Liaison Member) Professor of Mathematics, University of Minnesota at Minneapolis

R. Duncan Luce (Social and Behavioral Sciences Panel Liaison Member) Distinguished Professor of Cognitive Science and Director of the Irvine Research Unit in Mathematical Behavioral Science, University of California, Irvine

Ingram Olkin Professor of Statistics and Professor of Education, Stanford University

James Sethian Associate Professor of Mathematics, University of California, Berkeley

Audrey Terras (Biological and Health Sciences Panel Liaison Member) Professor of Mathematics, University of California, San Diego

P. Emery Thomas Professor of Mathematics, University of California, Berkeley

Panel Production Staff: Sara Wong, University of California, Berkeley

Physical and Information Sciences and Engineering

George Bugliarello (Panel Chair) President, Polytechnic University

James J. Conti Vice President for Educational Development, Polytechnic University

David H. Copp Member of the Technology Staff, Bell Communications Research

Kathleen Crane Associate Professor of Geology, Hunter College, City University of New York

Samuel Devons Professor Emeritus of Physics, Nevis Laboratories, Columbia University

Paul Garabedian Professor of Mathematics, Courant Institute of Mathematical Sciences, New York University

William Starnes   Professor of Chemistry, Polytechnic University

Alan Strahler   Professor of Geography, Boston University

Spencer Weart   Director, Center for History of Physics, American Institute of Physics

Panel Staff: Margaret Mastrianni, Polytechnic University

Social and Behavioral Sciences

Mortimer H. Appley   (Panel Chair)   Visiting Scholar in Psychology, Harvard University; President Emeritus, Clark University

Jill K. Conway   Visiting Scholar, Science, Technology, and Society, Massachusetts Institute of Technology; President Emeritus, Smith College

John T. Dunlop   Professor of Economics and Lamont University Professor Emeritus, Harvard University

Ann F. Friedlaender   Dean of Humanities, Massachusetts Institute of Technology

Jerome Kagan   Professor of Developmental Psychology, Harvard University

R. Duncan Luce   Distinguished Professor of Cognitive Science and Director of the Irvine Research Unit in Mathematical Behavioral Science, University of California, Irvine

Rosemarie Rogers   Professor of International Politics, The Fletcher School of Law and Diplomacy, Tufts University

Nur O. Yalman   Professor of Social Anthropology, Harvard University; Curator, Middle Eastern Ethnology, The Peabody Museum

Panel Staff: Winifred B. Maher, Lecturer, Harvard University Extension School

Technology

James R. Johnson   (Panel Chair)   Former Executive Scientist, 3M Company

Sister Marquita Barnard   Former Professor of Chemistry, The College of St. Catherine

Don Boyd   Director of Systems Technology and Engineering, Honeywell, Inc.

William Hamer   Vice President of Engineering, ADC Telecommunications

Robert T. Holt   Dean of the Graduate School, University of Minnesota at Minneapolis

Harvey Keynes   Professor of Mathematics, University of Minnesota at Minneapolis

John W. Pearson   Former Vice President of Development, 3M Company

Phillip Regal Professor of Ecology, University of Minnesota at Minneapolis

Matthew Tirrell Professor of Chemical Engineering and Materials Science, University of Minnesota at Minneapolis

Panel Staff: Karen Olson, Science Education Consultant (St. Paul, Minnesota)

PROJECT 2061 STAFF (AAAS)

F. James Rutherford Project Director; Chief Education Officer, American Association for the Advancement of Science

Andrew Ahlgren Associate Project Director; Professor of Education, University of Minnesota at Minneapolis

Patricia Warren Project Manager

Carol Holmes Secretary

Gwen McCutcheon Secretary

ADVISERS TO PROJECT 2061 STAFF

J. Myron Atkin Professor of Education, Center for Educational Research at Stanford, Stanford University

Michael Baker Account Supervisor, The Gallagher Widmeyer Group, Inc. (Washington, D.C.)

Ted Bartell U.S. Department of Education

James R. Beniger Associate Professor of Communications and Sociology, University of Southern California

Alfred Brown Former President, Celanese Corporation

Charles Brownstein Executive Officer, Directorate on Computer and Information Science and Engineering, National Science Foundation

Linda Campbell Consultant to National Education Association Mastery in Learning Project, Seattle Public Schools

William D. Carey Former Executive Officer, American Association for the Advancement of Science

Peggy Carnahan Secondary Science Supervisor, Northside Independent School District (San Antonio, Texas)

Donald L. Chambers Supervisor, Mathematics Education, Wisconsin Department of Public Instruction

Audrey B. Champagne Senior Program Director, American Association for the Advancement of Science

Bernard L. Charles Program Officer, Carnegie Corporation of New York

K. C. Cole Writer (New York City)

David Crandall Executive Director, The Network (Andover, Massachusetts)

Alden Dunham   Program Chair, Carnegie Corporation of New York

Karin D. Egan   Program Associate, Carnegie Corporation of New York

Paul M. Elliott   Wordsworth Communication (Alexandria, Virginia)

Joseph D. Exline   Associate Director of Science, Virginia Department of Education

Stuart I. Feldman   Division Manager, Bell Communications Research

Michael G. Fullan   Dean, Faculty of Education, University of Toronto

Stephen Gilbert   Vice President, New Programs, EDUCOM

Ronald L. Graham   Adjunct Director, AT&T Bell Laboratories

David A. Hamburg   President, Carnegie Corporation of New York

Johnnie Hamilton   Science Coordinator, Fairfax County Public Schools (Virginia)

David Hawkins   Professor Emeritus of Philosophy, University of Colorado, Boulder

Frances Hawkins   Cofounder, Mountain View Center, University of Colorado, Boulder

Patricia Heller   Assistant Professor of Curriculum and Instruction, University of Minnesota at Minneapolis

Elam Hertzler   Assistant Superintendent for Information Systems, Illinois State Board of Education

Shirley Hill   Professor of Education and Mathematics, University of Missouri at Kansas City

Cary C. Hoagland   Editorial Consultant, CCH Language Services (Takoma Park, Maryland)

Gerald Holton   Mallinckrodt Professor of Physics and Professor of History of Science, Harvard University

Lynn David Hubsch   Orion Studios (Washington, D.C.)

Susanne M. Humphrey   Information Scientist, National Library of Medicine

Paul DeHart Hurd   Professor Emeritus of Science Education, Stanford University

Arlene M. Kahn   Independent Educational Consultant, Arlene Kahn & Associates (New York City)

Mary L. Kiely   Program Associate, Carnegie Corporation of New York

Michael Kirst   Professor of Education, Stanford University

Manfred Kochen   Professor of Information Science, School of Medicine, University of Michigan

Laura Krich   Science Teacher, Diamond Middle School, Lexington, Massachusetts

Gerald Kulm   Project Director, American Association for the Advancement of Science

Douglas M. Lapp   Executive Director, National Science Resources Center, National Academy of Sciences/Smithsonian Institution

Charles LaRue   Coordinator of Elementary Science, Montgomery County Public Schools (Maryland)

Marvin Lazerson   Dean, School of Education, University of Pennsylvania

LeRoy Lee   Executive Director, Wisconsin Academy of Sciences, Arts and Letters

Michael E. Lesk   Manager, Computer Science Research Division, Bell Communications Research

Karen B. Levitan   President, The KBL Group, Inc. (Silver Spring, Maryland)

Shirley M. Malcom   Program Head, Office of Opportunities in Science, American Association for the Advancement of Science

Sheilah Mann   Director of Education, American Political Science Association

Robert M. McClure   Director, Mastery in Learning Project, National Education Association

Marge McClurg   Coordinator of Mathematics, Fairfax County Public Schools (Virginia)

Fay Metcalf   Executive Director, National Commission on Social Studies in the Schools

Wayne Moyer   Coordinator of Secondary Science, Montgomery County Public Schools (Maryland)

Robert F. Murray   Chief, Division of Medical Genetics, Department of Pediatrics, Howard University College of Medicine

Jimmy E. Nations   First-Grade Teacher, Westwood Primary School (Dalton, Georgia)

Fred Newmann   National Center on Effective Secondary Schools, University of Wisconsin–Madison

Elena Nightingale   Special Advisor to the President, Carnegie Corporation of New York

Michael O'Keefe   President, Consortium for the Advancement of Private Higher Education

Susie Oliphant   Assistant Director of Science, District of Columbia Public Schools

Lester G. Paldy   Director, Center for Science, Mathematics, and Technology Education, State University of New York at Stony Brook

Joan Palmer   Associate Superintendent for Curriculum and Supervision, Howard County Public Schools (Maryland)

Julie Phillips   Indexer (Vienna, Virginia)

Harold Pratt   Executive Director, Science and Technology Management, Jefferson County Public Schools (Colorado)

Diane Ravitch   Adjunct Professor of History and Education, Teachers College, Columbia University

David Z. Robinson   Executive Director, Carnegie Commission on Science, Technology, and Government

Bella Rosenberg   Assistant to the President, American Federation of Teachers

Kenneth Russell Roy   National Director, National Science Supervisors Association

Thomas Sachse   Manager, Mathematics and Science Education Unit, California Department of Education

Dan Saltrick   Director of Instruction, Prince George's County Public Schools (Maryland)

Ethel L. Schultz   Program Director, National Science Foundation

Cecily C. Selby   Professor of Science Education, New York University

Elliot R. Siegel   Assistant Director for Planning and Evaluation, National Library of Medicine

Elizabeth Stage   Director of Mathematics Education, Lawrence Hall of Science, University of California, Berkeley

Dorothy Stephens   Director of Instructional Services, District of Columbia Public Schools

James Strandquist   Director of Science, Prince George's County Public Schools (Maryland)

Lee Summerville   Executive Supervisor of Science, Howard County Public Schools (Maryland)

Marcia Sward   Executive Director, Mathematical Sciences Education Board, National Academy of Sciences

Constance Tate   Science Education Consultant (Washington, D.C.)

Chad A. Tolman   Research Leader, Central Research and Development Department, E. I. du Pont de Nemours and Company

Jerry Valadez   Science Teacher, Kings Kenyon Middle School (Fresno, California)

Decker Walker   Professor of Education, Stanford University

Fletcher G. Watson   Professor Emeritus of Science Education, Harvard University

Jeannette Wedel   Development Officer, American Association for the Advancement of Science

Wayne Welch   Professor of Educational Psychology, University of Minnesota at Minneapolis

Scott Widmeyer   Chairman, The Gallagher Widmeyer Group, Inc. (Washington, D.C.)

Marshall C. Yovits Professor of Computer and Information Science, Indiana University–Purdue University at Indianapolis

CONSULTANTS TO THE PHASE I PANELS

Biological and Health Sciences

Michael Criqui, M.D. Professor of Community and Family Medicine and Medicine, University of California, San Diego

Jared Diamond Professor of Physiology, University of California School of Medicine (Los Angeles)

Jean Lindsley Science Teacher, The Bishop's School (La Jolla, California)

D. William Rains Chairman, Department of Agronomy and Range Science, College of Agricultural and Environmental Sciences, University of California, Davis

Phoebe Roeder Lecturer, Department of Natural Sciences, San Diego State University

Mathematics

Zvonko Fazarinc Consulting Professor of Electrical Engineering, Stanford University; Director, University Relations, Hewlett–Packard Company

Lyle Fisher Codirector, Bay Area Mathematics Project, University of California School of Education (Berkeley)

William Kahan Professor of Computer Science, University of California, Berkeley

David Logothetti Associate Professor of Mathematics, Santa Clara University

Joel Schneider Content Director, Children's Television Workshop (New York City)

Alan Schoenfeld Chairman, Division of Education in Mathematics, Science and Technology, Graduate School of Education, University of California, Berkeley

Elizabeth Stage Director of Mathematics Education, Lawrence Hall of Science, University of California, Berkeley

Physical and Information Sciences and Engineering

Frank Aikman Associate Research Scientist, Lamont-Doherty Geological Observatory

Enrico Bonatti Senior Research Scientist, Lamont-Doherty Geological Observatory

Marvin P. Epstein Retired Associate, AT&T Bell Laboratories

Kenneth Ford Chief Executive Officer, American Institute of Physics

Cindy Lee   Associate Professor of Marine Environmental Studies, State University of New York at Stony Brook

David Mog   Senior Program Officer, Office of International Affairs, National Research Council

John Sanders   Professor of Geology, Barnard College

John G. Truxal   Distinguished Teaching Professor of Technology and Society, State University of New York at Stony Brook

Edward Wolf   Professor and Head of Physics Department, Polytechnic University

Social and Behavioral Sciences

Herbert P. Baker   Social Studies Teacher, Belmont High School (Belmont, Massachusetts)

Roger Brown   Chaired Professor of Psychology, Harvard University

Paul W. Carey   Head of Social Studies Department, Belmont High School (Belmont, Massachusetts)

Lily Gardner Feldman   Associate Professor of Political Science, Tufts University

Richard Herrnstein   Edgar Pierce Professor of Psychology, Harvard University

Jacob Irgang   Social Studies Teacher, Stuyvesant High School (New York City)

Brendan A. Maher   Chaired Professor of Psychology, Harvard University

Joseph S. Nye, Jr.   Professor of International Security, John F. Kennedy School of Government, Harvard University

Orlando Patterson   Professor of Sociology, Harvard University

David Riesman   Professor Emeritus of Sociology, Harvard University

Edward E. Smith   Professor of Psychology, University of Michigan

Robert M. Solow   Institute Professor of Economics, Massachusetts Institute of Technology

Janet Spence   Professor of Psychology, University of Texas at Austin

James Stellar   Professor of Psychology, Northeastern University

Sidney Verba   Professor of Political Science, Harvard University

Ezra F. Vogel   Chaired Professor of Sociology, Harvard University

Myron Weiner   Professor of Political Science, Massachusetts Institute of Technology

Technology

Laszlo A. Belady Vice President and Program Director, Software Technology Program, Microelectronics and Computer Technology Corporation (Austin, Texas)

M. James Bensen Dean, School of Industry and Technology, University of Wisconsin–Stout

Mario Bognanno Professor of Industrial Relations, University of Minnesota at Minneapolis

John Borchert Regents Professor of Geography, University of Minnesota at Minneapolis

Alfred B. Bortz Assistant Director, Magnetics Technology Center, Carnegie Mellon University

Kris K. Burhardt Vice President of Research and Development for the Information and Imaging Technologies Center, 3M Company

Elof Carlson Distinguished Teaching Professor of Biochemistry, State University of New York at Stony Brook

John Entorf Associate Dean of the School of Industry and Technology, University of Wisconsin–Stout

Robert M. Hexter Professor of Chemistry, University of Minnesota at Minneapolis

William F. Kaemmerer Fellow, Honeywell, Inc.

Jack Kilby Retired Senior Engineer, Texas Instruments

Robert Kudrle Professor of Public Relations, University of Minnesota at Minneapolis

Edwin Layton Professor of History of Science and Technology, University of Minnesota at Minneapolis

Douglas McCormick Editor in Chief, *Bio/Technology*

Mylon Eugene Merchant Director of Advanced Manufacturing Research, Metcut Research Associates, Inc. (Cincinnati, Ohio)

Jay Morgan Director of Research and Development, Pillsbury Company

Vernon W. Ruttan Regents Professor of Agricultural and Applied Economics, University of Minnesota at St. Paul

John Sadowski Assistant Professor of Electrical Engineering, Purdue University

Roger Staehle Consultant and Adjunct Professor of Chemical Engineering and Materials Science, University of Minnesota at Minneapolis

Sister Mary Thompson Chairperson of the Chemistry Department, College of St. Catherine

Paul Weiblen Professor of Geology, University of Minnesota at Minneapolis

Wendell Williams Professor and Chairman, Materials Science and Engineering Department, Case Western Reserve University

REVIEWERS OF *SCIENCE FOR ALL AMERICANS*

Walter H. Abelmann   Professor of Medicine, Harvard University

Allen F. Agnew   Courtesy Professor of Geology, Oregon State University

Björn Anderson   Department of Education and Educational Research, University of Göteborg (Sweden)

Richard D. Anderson   Boyd Professor Emeritus of Mathematics, Louisiana State University

Ronald D. Archer   Professor of Chemistry, University of Massachusetts–Amherst

Jane Armstrong   Senior Policy Analyst, Education Commission of the States

William F. Aspray   Associate Director, Charles Babbage Institute for the History of Information Processing, University of Minnesota at Minneapolis

J. Myron Atkin   Professor of Education, Center for Educational Research, Stanford University

Albert V. Baez   President, Vivamos Mejor (Greenbrae, California)

Adrienne Y. Bailey   Vice President for Academic Affairs, The College Board

Nancy S. Barrett   Professor of Economics and Chair, Department of Economics, The American University

John A. Bartley   Science Supervisor, Springfield School District (Springfield, Pennsylvania)

James T. Barufaldi   Director, Science Education Center, University of Texas at Austin

Robert Bensching   Director of Central Engineering, Gates Rubber Company

M. James Bensen   Dean, School of Industry and Technology, University of Wisconsin–Stout

Henry A. Bent   Professor of Chemistry, North Carolina State University at Raleigh

Joseph Beydler   Science Teacher, Horizon High School (Broomfield, Colorado)

Susan H. Bicknell   Professor of Forest Ecology, Humboldt State University

Paul J. Black   Professor and Head, Centre for Educational Studies, King's College, University of London, Chelsea Campus (Great Britain)

Henry Blackburn   Professor and Director, Division of Epidemiology, School of Public Health, University of Minnesota at Minneapolis

Rolf K. Blank   Director, Science and Mathematics Indicators Project, Council of Chief State School Officers

Walter Bodmer   Director of Research, Imperial Cancer Research Fund (Great Britain)

Juan A. Bonnet, Jr.   Director, Center for Energy and Environment Research, University of Puerto Rico, San Juan

Kenneth E. Boulding   Distinguished Professor Emeritus of Economics, and Research Associate and Project Director, Institute of Behavioral Science, University of Colorado, Boulder

Albert H. Boyd   Professor of Agronomy, Mississippi State University

Lewis M. Branscomb   Director of Science, Technology, and Public Policy, John F. Kennedy School of Government, Harvard University

Richard F. Brinckerhoff   Harlan Page Amen Professor and Instructor of Science, Phillips Exeter Academy

Stephen G. Brush   Professor of History of Science, Institute for Physical Science and Technology, University of Maryland at College Park

Alphonse Buccino   Dean, College of Education, University of Georgia

Thomas D. Cabot   Chairman Emeritus, Cabot Corporation

Bob Calcaterra   Director, Research and Development, Adolph Coors Company

Ann Card   Coordinator, Cadre Project, Colorado Department of Education

Iris M. Carl   Instructional Supervisor, Houston Independent School District (Texas)

Robert E. Chesley   Consultant in Education (Ojai, California)

Saul B. Cohen   President Emeritus, Queens College (New York City)

William L. Colville   Professor Emeritus of Agronomy, University of Georgia

Roy G. Creech   Professor and Head, Department of Agronomy, Mississippi State University

F. Joe Crosswhite   Professor of Mathematics, Northern Arizona University

James E. Davis   Chief Executive Officer, Instructional Design Associates (Boulder, Colorado)

John Dawson   Head of the Professional, Scientific, and International Affairs Division, British Medical Association

Linda R. DeTure   Director of Student Teaching, Rollins College

Thomas R. Dirksen   Associate Dean for Biological Sciences, Medical College of Georgia

John A. Dossey   Professor of Mathematics, Illinois State University

James L. Doud   Professor of School Administration, Department of Educational Administration and Counseling, University of Northern Iowa

Rosalind Driver School of Education, Leeds University (Great Britain)

Daniel C. Drucker Graduate Research Professor, College of Engineering, University of Florida

J. R. Durant Department of External Studies, University of Oxford (Great Britain)

Harriet P. Dustan Professor Emeritus of Medicine, University of Alabama in Birmingham

Annagreta Dyring Forskningsrådsnämnden, Swedish Council for Planning and Coordination of Research

Jack Easley Professor of Teacher Education, University of Illinois, Urbana-Champaign

James L. Elder President, The School for Field Studies (Beverly, Massachusetts)

Yehuda Elkana Van Leer Jerusalem Institute (Israel)

Jerry Elliott Technical Manager for Support Equipment, Space Station Program, Johnson Space Center

John D. Emerson Dean of the College, Middlebury College

James Fey Professor of Curriculum and Instruction and Mathematics, University of Maryland at College Park

Richard I. Ford Associate Dean for Research and Computing, University of Michigan

William R. Freudenburg Associate Professor of Rural Sociology, University of Wisconsin–Madison

Alice B. Fulton Associate Professor of Biochemistry, University of Iowa

Charles O. Gardner Foundation Professor, Department of Agronomy, University of Nebraska at Lincoln

Ronald Geballe Professor Emeritus of Physics, University of Washington

B. Frank Gillette Former Superintendent of Schools, Los Gatos-Saratoga (California)

Owen Gingerich Professor of Astronomy and History of Science, Harvard–Smithsonian Center for Astrophysics

Steven Goldman Professor of Philosophy and History, and Andrew W. Mellon Professor of Humanities, Lehigh University

Mary L. Good Senior Vice President, Technology, Allied-Signal Inc. (Morristown, New Jersey)

Thomas A. Gorell Associate Dean and Professor of Biology, College of Natural Sciences, Colorado State University

Judith P. Grassle Senior Scientist, Marine Biological Laboratory (Woods Hole, Massachusetts)

Richard Gregory Brain and Perception Laboratory, The Medical School (Bristol, Great Britain)

Richard J. Griego Professor of Mathematics, University of New Mexico

Norman Hackerman   President Emeritus, William Marsh Rice University

Sheila M. Haggis   Chief, Science Education Section, Division of Science, Technical, and Environmental Education, United Nations Educational, Scientific and Cultural Organization (Paris, France)

Archibald O. Haller   Professor of Rural Sociology, University of Wisconsin, Madison

William W. Hambleton   Professor Emeritus of Geology, University of Kansas

John Harris   Centre for Educational Studies, King's College, University of London, Chelsea Campus (Great Britain)

Anna J. Harrison   Professor Emeritus of Chemistry, Mount Holyoke College

Bernt Hauge   Ugla School (Trondheim, Norway)

Henry Heikkinen   Professor of Chemistry, University of Northern Colorado

Robert L. Heller   Chancellor Emeritus, University of Minnesota at Duluth

Ernest M. Henley   Professor of Physics, University of Washington

Robert L. Hirsch   Vice President, Research and Technical Services, ARCO Oil and Gas Company

Betty-ann Hoener   Associate Professor of Pharmacy and Pharmaceutical Chemistry, University of California, San Francisco

Donald F. Holcomb   Professor of Physics, Cornell University

John E. Hopcroft   Chairman, Department of Computer Sciences, and Joseph C. Ford Professor of Computer Science, Cornell University

James H. Hubbard   Associate Director, Colorado Alliance for Science

Paul DeHart Hurd   Professor Emeritus of Science Education, Stanford University

Daphne Jackson   Department of Physics, University of Surrey (Guilford, Great Britain)

David Jenness   Scholar in Residence, National Commission on Social Studies in the Schools

Francis S. Johnson   Professor of Physics, University of Texas at Dallas

Larry Johnson   Dean, School of Letters, Arts, and Science, Metropolitan State College

Richard A. Y. Jones   Dean, School of Chemical Sciences, University of East Anglia (Norwich, Great Britain)

Irene Jordan   Chemistry Teacher, Greeley Central High School (Greeley, Colorado)

Jane Butler Kahle   Professor of Biological Sciences and Education, Purdue University

Ann Kahn Past President, National Parent Teacher Association

Anna Karlin Principal Software Engineer, Digital Equipment Corporation (Palo Alto, California)

J. Katus Subfaculteit der Pedagogische en Andragogische Wetenschappen, University of Leiden (the Netherlands)

Alice B. Kehoe Professor of Anthropology, Marquette University

Manert H. Kennedy Executive Director, Colorado Alliance for Science

Judith T. Kildow Associate Professor of Ocean Policy, Massachusetts Institute of Technology

Philip Kitcher Professor of Philosophy, University of California, San Diego

Sally Gregory Kohlstedt Professor of History, Syracuse University

Gretchen S. Kolsrud Manager, Biological Applications Program, Office of Technology Assessment, U.S. Congress

Gerald Kulm Project Director, American Association for the Advancement of Science

James M. Landwehr Supervisor, Statistical Models and Methods Research Department, AT&T Bell Laboratories

Michael Lang State Science Supervisor, Arizona State Department of Education

Edwin T. Layton, Jr. Professor of History of Science and Technology, University of Minnesota at Minneapolis

Richard M. Lemmon Senior Chemist Emeritus, Lawrence Berkeley Laboratory

Simon A. Levin Charles A. Alexander Professor of Biological Sciences, and Director, Center for Environmental Research, Cornell University

Robert J. Levine Professor of Medicine, Yale University

Alan P. Lightman Professor of Science and Writing, Massachusetts Institute of Technology

John A. Limongiello Science Teacher, Winchester High School (Winchester, Massachusetts)

Harald Löe Director of National Institute of Dental Research, National Institutes of Health

Gil Lopez Director, Comprehensive Math and Science Program, Columbia University

Vincent N. Lunetta Associate Dean for Research and Graduate Studies, Pennsylvania State University

Shirley M. Malcom Program Head, Office of Opportunities in Science, American Association for the Advancement of Science

Sheilah Mann Director of Education, American Political Science Association

Nancy H. Marcus Associate Professor of Oceanography, Florida State University

Stephanie Pace Marshall Director, Illinois Math and Science Academy

Paul C. Martin Dean, Division of Applied Sciences, Harvard University

Jessica Tuchman Mathews Vice President and Director of Research, World Resources Institute

John Matis Computer Analyst, Bureau of Land Management

John May Centre for Science Education, King's College, University of London, Chelsea Campus (Great Britain)

Victor J. Mayer Professor of Science Education, Ohio State University

James J. McCarthy Alexander Agassiz Professor of Biological Oceanography, Harvard University

Joseph D. McInerney Director, Biological Sciences Curriculum Study

Blaine C. McKusick Assistant Director Emeritus, Haskell Laboratory of Toxicology, E. I. du Pont de Nemours and Company

Barbara R. Migeon Director, Predoctoral Program in Human Genetics and Molecular Biology, The Johns Hopkins University

Jon D. Miller Professor of Political Science, The Graduate School, and Director, Public Opinion Laboratory, Northern Illinois University

Robert Moore Professor of Sociology, University of Aberdeen (Scotland)

Brian Oakley Secretary Emeritus, The Science and Engineering Research Council (Great Britain)

Robert H. Page Forsyth Professor of Mechanical Engineering, Texas A&M University

Jay M. Pasachoff Field Memorial Professor of Astronomy, Williams College

Naomi Pasachoff Williams College

Marjorie L. Reaka Associate Professor of Zoology, University of Maryland at College Park

Nat C. Robertson Director and Science Advisor, Marion Laboratories, Inc. (Kansas City, Missouri)

Thomas A. Romberg Professor of Curriculum and Instruction, University of Wisconsin–Madison

Walter G. Rosen Senior Program Officer, Board on Basic Biology, National Research Council

Mary Budd Rowe Professor of Science Education, University of Florida

Kenneth Russell Roy National Director, National Science Supervisors Association

Rustum Roy   Director, Science, Technology and Society Program, Pennsylvania State University

J. Oliver Ryan   Department of Education, University College of Galway, National University of Ireland

Donald O. Schneider   Professor and Department Head, Social Science Education Department, University of Georgia

Frederick Seitz   President Emeritus, Rockefeller University

Cecily C. Selby   Professor of Science Education, New York University

Lawrence Senesh   Professor Emeritus of Economics, University of Colorado, Boulder

Sydel Silverman   President, Wenner-Gren Foundation for Anthropological Research, Inc.

Derek Smyth   (Shoreham-by-Sea, Great Britain)

Colin Spedding   Pro-Vice Chancellor, University of Reading (Great Britain)

Kendall N. Starkweather   Executive Director, International Technology Education Association

Lynn A. Steen   Professor of Mathematics, St. Olaf College

Frank J. Sulloway   MacArthur Fellow, Harvard University

John Swets   Chief Scientist, BBN Laboratories Inc. (a subsidiary of Bolt Beranek and Newman Inc.) (Cambridge, Massachusetts)

M. W. Thring   Professor Emeritus of Chemical Engineering, Queen Mary College (London, Great Britain)

Charles R. Tolbert   Associate Professor of Astronomy, University of Virginia

James W. Trott, Jr.   Director, School of Vocational Education, Louisiana State University and A&M College

John G. Truxal   Distinguished Teaching Professor of Technology and Society, State University of New York at Stony Brook

John W. Tukey   Professor Emeritus of Statistics, Princeton University

Ruth D. Turner   Professor of Biology and Curator in Malacology at the Museum of Comparative Zoology, Harvard University

Anne Tweed   Biology Teacher, Smoky Hill High School (Aurora, Colorado)

Bruce L. Umminger   Acting Director, Division of Cellular Biosciences, National Science Foundation

Burton E. Voss   Professor of Science Education, University of Michigan

Herbert J. Walberg   Research Professor of Education, University of Illinois, Chicago

Sylvia A. Ware   Director of Education, American Chemical Society

Ed Waterman   Chemistry Teacher, Rocky Mountain High School (Fort Collins, Colorado)

Richard L. Webber   Chief of Diagnostic Systems Branch, National Institute of Dental Research, National Institutes of Health

Gunter Weller   Professor of Geophysics, University of Alaska, Fairbanks

Edward Wenk, Jr.   Professor Emeritus of Engineering and Public Affairs, University of Washington

F. Karl Willenbrock   Executive Director, American Society for Engineering Education

H. S. Wolff   Brunel Institute of Bio-Engineering, Brunel University (Great Britain)

Dael L. Wolfle   Professor Emeritus of Public Affairs, Graduate School of Public Affairs, University of Washington

Robert E. Yager   Professor of Science Education, University of Iowa

Chris Zafiratos   Associate Vice Chancellor, University of Colorado, Boulder

John Ziman   Professor Emeritus of Theoretical Physics, Bristol University (Great Britain)

REVIEWERS OF PHASE I PANEL REPORTS

Dean Abrahamson   Professor, Hubert H. Humphrey Institute of Public Affairs, University of Minnesota at Minneapolis

Guenther Albrecht-Buehler   Associate Full Professor of Cell Biology and Anatomy, Northwestern University

Richard D. Anderson   Boyd Professor Emeritus of Mathematics, Louisiana State University and A&M College

Nancy S. Barrett   Professor of Economics and Chair, Department of Economics, The American University

Henry Blackburn   Professor and Director, Division of Epidemiology, School of Public Health, University of Minnesota at Minneapolis

Joseph Bordogna   Dean, School of Engineering and Applied Science, University of Pennsylvania

Kathryn Borman   Associate Dean of Graduate Studies and Research, and Professor of Education and Sociology, College of Education, University of Cincinnati

Audrey Brainard   Elementary Science Consultant (Holmdel, New Jersey)

Alphonse Buccino   Dean, College of Education, University of Georgia

Gail Burrill   Mathematics Teacher, Whitnall High School (Greenfield, Wisconsin)

Rodger Bybee   Associate Director, Biological Sciences Curriculum Study

Rosemary Chalk   Former Program Head, Office of Scientific Freedom and Responsibility, American Association for the Advancement of Science

Saul B. Cohen   President Emeritus, Queens College (New York City)

R. Kent Crawford   Physicist, Argonne National Laboratory

Roy G. Creech   Professor and Head, Department of Agronomy, Mississippi State University

William V. D'Antonio   Executive Officer, American Sociological Association

Robert B. Davis   New Jersey Professor of Mathematics Education, Rutgers–The State University

John A. Dossey   Professor of Mathematics, Illinois State University

Kenneth Ford   Chief Executive Officer, American Institute of Physics

Charles O. Gardner   Foundation Professor, Department of Agronomy, University of Nebraska

John E. Gaustad   Professor of Astronomy, Swarthmore College

Gary D. Glenn   Associate Professor of Political Science, Northern Illinois University

Steven Goldman   Professor of Philosophy and History, and Andrew W. Mellon Professor of Humanities, Lehigh University

Harold Green   Professor of Law, George Washington University National Law Center

Janice G. Hamrin   Executive Director, Independent Energy Producers Association

Jon Harkness   K–12 Science Specialist, Wausau School District (Wausau, Wisconsin)

Anna J. Harrison   Professor Emeritus of Chemistry, Mount Holyoke College

Robert Highsmith   Director of Research, Joint Council on Economic Education

Donald F. Holcomb   Professor of Physics, Cornell University

David D. Houghton   Professor of Meteorology, University of Wisconsin–Madison

L. Douglas James   Director, Utah Water Research Laboratory, Utah State University

Neal F. Johnson   Professor of Psychology, Ohio State University

Jane Butler Kahle   Professor of Biological Sciences and Education, Purdue University

Marvin E. Kauffman   Executive Director, American Geological Institute

Judith T. Kildow   Associate Professor of Ocean Policy, Massachusetts Institute of Technology

Gregory A. Kimble Professor Emeritus of Psychology, Duke University

Gretchen S. Kolsrud Manager, Biological Applications Program, Office of Technology Assessment, U.S. Congress

James M. Landwehr Supervisor, Statistical Models and Methods Research Department, AT&T Bell Laboratories

Anneli Lax Professor of Mathematics, New York University

Margaret D. LeCompte Writer and Educational Researcher (Houston, Texas)

Fred Leone Executive Director, American Statistical Association

Karen B. Levitan President, The KBL Group, Inc. (Silver Spring, Maryland)

Vincent N. Lunetta Associate Dean for Research and Graduate Studies, Pennsylvania State University

Peter Lykos Professor of Chemistry, Illinois Institute of Technology

Shirley M. Malcom Program Head, Office of Opportunities in Science, American Association for the Advancement of Science

James Marran Chairman, Social Studies Department, New Trier High School (Winnetka, Illinois)

Robert B. McCall Director, Office of Child Development, University of Pittsburgh

Helen M. McCammon Director, Ecological Research Division, U.S. Department of Energy

Henry A. McGee, Jr. Professor of Chemical Engineering, Virginia Polytechnic Institute and State University

Blaine C. McKusick Assistant Director Emeritus, Haskell Laboratory of Toxicology, E. I. du Pont de Nemours and Company

Barbara R. Migeon Director, Predoctoral Program in Human Genetics and Molecular Biology, The Johns Hopkins University

David Nanney Professor of Ecology, Ethology and Evolution, University of Illinois, Urbana-Champaign

Robert H. Page Forsyth Professor of Mechanical Engineering, Texas A&M University

Jay M. Pasachoff Field Memorial Professor of Astronomy, Williams College

Michael I. Posner Professor of Psychology, University of Oregon

William F. Prokasy Vice President of Academic Affairs, University of Georgia

Ronald Pyke Professor of Mathematics, University of Washington

Anthony Ralston Professor of Computer Science and Mathematics, State University of New York at Buffalo

John S. Rigden Professor of Physics, University of Missouri at St. Louis

Rustum Roy Director, Science, Technology and Society Program, Pennsylvania State University

Jerome Sacks Head, Department of Statistics, University of Illinois, Urbana-Champaign

Paul Sally Professor of Mathematics, University of Chicago

Theodore Schlie Associate Professor of Business Administration, Illinois Institute of Technology

Willis Sibley Professor of Anthropology, Cleveland State University

Sidney Simpson Professor and Head, Department of Biological Sciences, University of Illinois, Chicago

Jeremiah Stamler Dingman Professor of Cardiology, Northwestern University

Lynn A. Steen Professor of Mathematics, St. Olaf College

Charles R. Tolbert Associate Professor of Astronomy, University of Virginia

Ruth D. Turner Professor of Biology and Curator in Malacology at the Museum of Comparative Zoology, Harvard University

Zalman Usiskin Professor of Education, University of Chicago

Sylvia A. Ware Director of Education, American Chemical Society

Robert E. Yager Professor of Science Education, University of Iowa

APPENDIX B
SELECTED REFERENCES

• Adler, Mortimer J. 1982. *The Paideia Proposal*. New York: Macmillan Publishing. (Assumes as a basic premise that all students should receive the same fundamental education. Proposes a rigorous curriculum that stresses acquisition of organized knowledge, development of intellectual skills, and enlarged understanding of ideas and values. Also addresses roles of teachers and administrators in this proposed education system.)

• Boyer, Ernest L. 1983. *High School: A Report on Secondary Education in America*. New York: Harper & Row. (Presents a detailed agenda for national reform in secondary education. Identifies twelve key strategies for achieving high-quality education, including a high school core curriculum that calls for two years of mathematics and a semester in which students study the impact of technology. Funded by the Carnegie Foundation for the Advancement of Teaching.)

• Business–Higher Education Forum. 1983. *America's Competitive Challenge: The Need for a National Response*. Washington, D.C.: The Forum. (Identifies the symptoms of America's economic decline and describes the contributing factors. Offers recommendations on trade, capital investment, private-sector initiatives, human resources, and technological innovation. Also suggests initiatives for businesses and colleges, as well as ways they can improve their own operating procedures.)

• Carnegie Corporation of New York. 1983. *Education and Economic Progress. Toward a National Education Policy: The Federal Role*. New York: The Corporation. (Recommends ways in which education policy and reform can serve to strengthen America's increasingly science and technology-based economy. Focuses largely on the need for heightened general scientific awareness and scientific involvement in curriculum development. Offers recommendations regarding vocational training; the federal, state, and local roles in reforming education; servicing underserved students; and teaching.)

• Carnegie Forum on Education and the Economy, Task Force on Teaching as a Profession. 1986. *A Nation Prepared: Teachers for the 21st Century*. New York: The Forum. (Makes the case for the need for drastic reform in the quality and evaluation of teaching in American schools. Offers a reform plan for raising and broadening standards of eligibility and performance for teachers, restructuring schools to enable teachers to teach more effectively, and increasing incentives and rewards for teachers.)

• Center for the Assessment of Educational Progress, Educational Testing Service. 1987. *The Subtle Danger: Reflections on the Literacy Abilities of America's Young Adults*. Princeton, N.J.: Educational Testing Service. (Examines the implications of ETS's assessment of literacy in America. Focuses on the effects of widespread illiteracy on employment, national economic competitiveness, poverty, and inequality among subgroups. Includes an evaluation of literacy skills in America and of the cognitive and literacy demands of different kinds of texts.)

• Champagne, Audrey B., ed. 1988. *Science Teaching: Making the System Work*. Washington, D.C.: American Association for the Advancement of Science. (A volume in an ongoing series entitled This Year in School Science. The series chronicles the development of issues and ideas related to each of three major elements of science education—science teachers, the science curriculum, and the learner. Each volume contains up-to-date statistics and essays by leaders in the field analyzing the issues and critiquing proposed public policy initiatives to advance science education. The other volumes, to date, are *Students and Science Learning* [1987], *The Science Curriculum* [1986], and *Science Teaching* [1985].)

• Education Commission of the States. 1984. *Action in the States: Progress Toward Education Renewal*. Denver: Gordon Bell Press. (Report of the Task Force on Education for Economic Growth. Reviews state programs, state legislation, and state board decisions that affect U.S. elementary and secondary education. Discusses successful state initiatives and offers recommendations on such topics as state plans, partnerships, school leadership, teaching, and underserved students. Also identifies promising initiatives for improving science, mathematics, computer, and technology education.)

• Exxon Education Foundation. 1984. *Science Education in the United States: Essential Steps for Achieving Fundamental Improvement*. New York: The Foundation. (Explores some of the major issues raised by the science education reform movement. Offers recommendations as to what steps must be taken to ensure implementation of improvements for elementary and secondary science education. Cites the need for a new conceptual framework for science and technology education; discusses the goals, curricula, and materials for science education; and focuses on the training of science teachers, research, and business and science partnerships.)

• Goodlad, John I. 1983. *A Place Called School: Prospects for the Future*. New York: McGraw-Hill. (Reports on the findings from A Study of Schooling, a five-year inquiry into U.S. secondary education. Focuses on how schools measure up against stated goals for education, how schools could be improved, how the individual school is the key unit for change, and what kinds of data a school should collect about itself to begin the improvement process.)

• The Holmes Group. 1986. *Tomorrow's Teachers: A Report of the Holmes Group*. East Lansing, Mich.: The Group. (Report of a consortium of deans and chief academic officers from research institutions in each U.S. state. Presents a broad agenda for improving the quality of America's teachers. Recommendations focus on certification, recruitment, standards, accountability, evaluation, and teaching environments.)

• International Association for the Evaluation of Educational Achievement. 1987. *The Underachieving Curriculum: A National Report on the Second International Mathematics Study*. Champaign, Ill.: Stipes Publishing. (Based on the second international study of primary and secondary mathematics. Assesses the quality of U.S. mathematics education. Discusses the content of curricula, quality of teaching, and students' achievement and attitudes.)

• International Association for the Evaluation of Educational Achievement. 1988. *Science Achievement in Seventeen Countries: A Preliminary Report*. Oxford: Pergamon Press. (Presents initial findings from IEA's second international study of science achievement (1983–86). Discusses the results for three school population levels. Also considers the growth in achievement between population levels, and at sex differences in science achievement.)

• Klein, Margrete S., and F. James Rutherford, eds. 1985. *Science Education in Global Perspective: Lessons From Five Countries.* Boulder, Colo.: Westview Press. (Examines the educational systems of Japan, China, East and West Germany, and the Soviet Union, which have developed particularly innovative approaches to science education. Provides an international cross section of useful comparative data.)

• Mullis, Ina V. S., and Lynn B. Jenkins. 1988. *The Science Report Card, Elements of Risk and Discovery: Trends and Achievement Based on the 1986 National Assessment.* Princeton, New Jersey: Educational Testing Service. (Reports trends over the last 20 years in data from the National Assessment of Educational Progress on science knowledge of students in grades 3, 7, and 11. Also attempts to correlate knowledge with exposure to science instruction, teachers' qualifications, classroom practices, parental involvement, and extracurricular experiences. An interpretive overview by Richard Berry, Audrey Champagne, John Pennick, Senta Raizen, Iris Weiss, and Wayne Welch puts the results into coherent perspective.)

• National Commission on Secondary Vocational Education. 1984. *The Unfinished Agenda: The Role of Vocational Education in the High School.* Washington, D.C.: The Commission. (Explores the role and function of U.S. secondary vocational education. Covers such issues as perceptions of vocational education, recruitment and preparation of teachers, standards and accountability, and partnerships involving schools, business, labor, and the community at large. Recommendations include a flexible curriculum across academic and vocational areas and a plan for recruiting teachers.)

• The National Commission on Excellence in Education. 1983. *A Nation at Risk: The Imperative for Educational Reform.* Washington, D.C.: U.S. Department of Education. (A major report that declares that the U.S. educational system has failed to meet the nation's needs. Describes deficiencies of U.S. high school education and lack of achievement among students. Presents recommendations for reform in curriculum content, standards and expectations, time devoted to schoolwork, teaching, leadership, and financial support.)

• National Governors' Association. 1987. *Making America Work. Bringing Down the Barriers. Productive People Productive Policies.* Washington, D.C.: The Association. (Presents the policy recommendations of task forces of governors on welfare prevention, school dropouts, teen pregnancy, adult literacy, and alcohol and drug abuse. Also focuses on policy changes in each of these areas that would make America more competitive economically.)

• National Governors' Association, Center for Policy Research and Analysis. 1986. *Time for Results: The Governors' 1991 Report on Education.* Washington, D.C.: The Association. (Presents governors' responses to seven key questions for education reform. Establishes a policy approach to teachers' accountability, school leadership, vouchers, enrichment and remedial programs, school-year length, instructional technologies and resources, and college students' achievement.)

• National Science Board Commission on Precollege Education in Mathematics, Science, and Technology. 1983. *Educating Americans for the 21st Century: A Report to the American People and the National Science Board*. Washington, D.C.: National Science Foundation. (Presents a detailed strategy for making U.S. mathematics, science, and technology education the world's best by 1995. Proposes major changes in the breadth of students' participation, methods and quality of teaching, preparation and motivation of students, and standards of achievement for students. Also details financial and government procedural changes that would be involved in the proposed improvements.)

• National Science Foundation and the U.S. Department of Education. 1980. *Science and Engineering Education for the 1980's and Beyond*. Washington, D.C.: The Foundation. (Prepared as a response to President Carter's concerns about the adequacy of science and engineering education for America's long-term needs. Documents the deficiencies and general decline of U.S. science and engineering education at secondary and postsecondary levels. Details the consequence of that decline for the scientific community and for federal policy.)

• Panel on Secondary School Education for the Changing Workplace. 1984. *High Schools and the Changing Workplace: The Employers' View*. Washington, D.C.: National Academy Press. (Focuses on the needs of high school graduates entering the labor force. Describes a set of core competencies—knowledge, skills, attitudes, and habits—that employers believe will equip young people for success in the labor market. Prepared by a panel formed under the auspices of a joint committee of the National Academy of Sciences, the National Academy of Engineering, and the Institute of Medicine.)

• Powell, Arthur G., Eleanor Farrar, and David K. Cohen. 1985. *The Shopping Mall High School: Winners and Losers in the Educational Marketplace*. Boston: Houghton Mifflin. (Describes how high schools try to offer something educationally rewarding for many different kinds of students and to graduate as many students as possible—and how they necessarily ignore the needs of the students who make few or no educational demands; that is, the average students who graduate with only a mediocre education. One of several books based on the five-year inquiry into secondary education entitled A Study of High Schools.)

• "Scientific Literacy" (entire issue). 1983. *Daedalus* (Spring). Issued as vol. 112, no. 2 of the Proceedings of the American Academy of Arts and Sciences. (Eleven essays on various aspects of scientific literacy.)

• Task Force on Education for Economic Growth. 1983. *ACTION for Excellence: A Comprehensive Plan to Improve Our Nation's Schools*. Denver: Hirschfeld Press. (Offers recommendations for improving U.S. education by clarifying the goals of education and establishing the means to realize these goals. Stresses need for action at the state and local level to involve community and business elements, to marshal resources effectively, to strengthen and reward teachers of higher quality, to improve school leadership, and to serve students equitably. Also describes in general the mathematical, scientific, computer, and reasoning skills that every student should develop.)

• Twentieth Century Fund Task Force on Federal Elementary and Secondary Education Policy. 1983. *Making the Grade*. New York: Twentieth Century Fund. (Presents recommendations on the proper role of the federal government in promoting high-quality education within the context of state and local control. Recommends that the federal government emphasize programs to develop basic scientific literacy among all citizens and to provide advanced training in sciences and mathematics for secondary students.)

INDEX
AND CONTENT
GUIDE

INDEX AND CONTENT GUIDE

This section consists of both an index to the entire report and a content guide to the recommendations of the National Council on Science and Technology Education presented in Part II. The content guide is made up of individual sets of entries positioned alphabetically within the index. Each such set is headed by a section title (printed in large and small capital letters) and page numbers, and it contains a list of the topics covered in the section.

as a distinguishing factor between humans and other species, 67–68
history of scientific endeavor and, 111
and perceptions about mental health, 73
scientific values and attitudes and, 133–35
technology and, 39, 67
transition from hunter-gatherer to farmer, 68, 90
See also Behavior, human
Curie, Marie and Pierre, 116
Curriculum, 4–5, 14, 20–21, 161–62
Cycles
  biological, 61
  of chemical elements on earth, 50, 63
  of decay, organic, 63
  of earth changes, 114–15
  in ecosystem fluctuations, 62, 63
  energy, in cells, 63
  food webs, 60, 62–63, 67, 90
  human life, 68–70
  matter and energy, 62–63
  as pattern of change, 128
  of social change, 128
  in systems, 129
  in vibration, 128
  water, and climate, 49
  weather, 128

Dalton, John, 116
Darwin, Charles, 111, 114, 117–18, 134
Data collection
  through manipulation and observation, 137–38
  sped up by technology, 29
  and technology decisions, 45
  See also Statistical analysis
Decay
  cycle for organisms, 63
  isotopic, 52, 117, 128
Decision-making processes
  in capitalist systems, 83
  computer applications in, 97
  distance effect in, 82
  for peaceful social change, 81
  prediction of consequences of, 86
Deforestation, 50, 85, 90, 92, 94
Design, 41
DNA
  deletion, insertion, or substitution of segments, 61
  gene sequence, 60, 90, 118
  instructions for protein assembly, 5, 61
  location in cells, 61
  molecular structure, 51, 60, 118

Earth
  THE EARTH (section), 48–50
  Sequence of topics:
    composition and motion
    climates
    mineral cycles
    rock cycle and strata
    effects of life

FORCES THAT SHAPE THE EARTH (section), 50
  Sequence of topics:
    interior and moving plates
    weathering
    mineral cycles
    rock cycle and strata
    effects of life
SETTING THE EARTH'S SURFACE IN MOTION (section), 115
  Sequence of topics:
    discovering matches between continents
    theory of moving crustal plates
    plate tectonics as a unifying principle
EXTENDING TIME (section), 114–15
  Sequence of topics:
    old ideas and evidence about the age of the earth
    argument for a very old earth
    slow cycles of change versus sometimes abrupt evolution
DISPLACING THE EARTH FROM THE CENTER OF THE UNIVERSE (section), 112–13
  Sequence of topics:
    ancient model of circles around earth
    circles or elipses around the sun
    telescopic observations
    resistance from clergy
  age and composition, 47, 48, 114–15, 134
  atmosphere, 48–49
  change processes, 114
  climatic patterns, causes of, 49
  distance from sun, 49
  evolution of life on, 48
  fresh water, 49
  gravitational force, 48
  interior, 48, 50, 54–55
  magnetic field, 56
  natural resources, 49
  representation on flat maps, 104
  rock sediment record, 50
  seasonal variations, cause of, 49
  tides, 48, 123
  water, forms on, 49
Earthquakes
  cause of, 27, 115
  patterns of, 128
  prediction of, 36, 104, 128
  vibratory motion from, 54, 128
Ecosystems
  biomass, constancy of, 63
  causes of changes in, 62
  cyclic fluctuations in, 62, 63
  and engineering design, 41
  energy flows in, 63, 123
  equilibrium in, 127
  exotic species introduced into, 62, 90, 91
  feedback in, 124
  forces shaping, 62
  habitat destruction in, 90
  interdependence of organisms in, 62
  stability of, 62, 64, 127
  variety of, 62
  See also Living organisms
Economics
  in engineering design, 41

ILLUSTRATION CREDITS: p. 10—from *Bulletin de la Société Astronomique* (1910), reprinted in *The Comet Book,* by Robert D. Chapman and John C. Brandt (Boston and Portola Valley, Calif.: Jones and Bartlett Publishers, 1984); p. 18—photo courtesy Dennis Mammana, San Diego, Calif.; p. 24—photo courtesy the Collection of the Whitney Museum of American Art, Mrs. John B. Putnam Bequest, acq. no. 84.41; p. 32—from *Science,* vol. 170 (October 23, 1970); p. 38—from *Science,* vol. 197 (August 26, 1977); p. 46—photo courtesy the National Gallery of Art, Washington, D.C., acc. no. 2678; p. 58—photo courtesy the National Gallery of Art, Washington, D.C., John Hay Whitney Collection, acc. no. 1982.76.7; p. 66—photo courtesy the Art Museum, Princeton University, acc. no. 67-2; p. 76—photo courtesy the National Gallery of Art, Washington, D.C., Paul Mellon Collection, acc. no. 1965.16.4; p. 88—photo courtesy the Minneapolis Institute of Arts, acc. no. 80.28; p. 100—photo courtesy the Hirshhorn Museum and Sculpture Garden, Smithsonian Institution, gift of the Joseph H. Hirshhorn Foundation, 1966 (Davis); p. 110—photos courtesy the University of Oklahoma Library and the Harvard College Observatory; p. 122—photo courtesy The Textile Museum, Washington, D.C., acc. no. R36.14.2B; p. 132—from *Leonardo da Vinci* (New York: Reynal & Co., 1956); p. 142—photo courtesy Rogier Windhorst and Alan Dressler, Carnegie Institution of Washington; p. 144—photo courtesy the National Gallery of Art, Washington, D.C., Andrew W. Mellon Collection, acc. no. 1937.1.91; p. 154—photo courtesy the National Gallery of Art, Washington, D.C., acc. no. 1979.13.3; p. 160—photo courtesy the National Gallery of Art, Washington, D.C., gift of Mr. C. V. S. Roosevelt (1974), acc. no. B-26,963.

Book Design: Orion Studios, Washington, D.C.